The STEAM Revolution

AF213191

Armida de la Garza • Charles Travis
Editors

The STEAM Revolution

Transdisciplinary Approaches to Science, Technology, Engineering, Arts, Humanities and Mathematics

Springer

IRISH RESEARCH COUNCIL
An Chomhairle um Thaighde in Éirinn

Editors
Armida de la Garza
College of Arts, Celtic Studies and Social Sciences
University College Cork
Cork, Ireland

Charles Travis
Department of History
University of Texas at Arlington
Arlington, TX, USA

Trinity Centre for Environmental Humanities
Trinity College Dublin
Dublin, Ireland

ISBN 978-3-030-07867-6 ISBN 978-3-319-89818-6 (eBook)
https://doi.org/10.1007/978-3-319-89818-6

© Springer International Publishing AG, part of Springer Nature 2019
Softcover re-print of the Hardcover 1st edition 2019
This work is subject to copyright. All rights are reserved by the Publisher, whether the whole or part of the material is concerned, specifically the rights of translation, reprinting, reuse of illustrations, recitation, broadcasting, reproduction on microfilms or in any other physical way, and transmission or information storage and retrieval, electronic adaptation, computer software, or by similar or dissimilar methodology now known or hereafter developed.
The use of general descriptive names, registered names, trademarks, service marks, etc. in this publication does not imply, even in the absence of a specific statement, that such names are exempt from the relevant protective laws and regulations and therefore free for general use.
The publisher, the authors and the editors are safe to assume that the advice and information in this book are believed to be true and accurate at the date of publication. Neither the publisher nor the authors or the editors give a warranty, express or implied, with respect to the material contained herein or for any errors or omissions that may have been made. The publisher remains neutral with regard to jurisdictional claims in published maps and institutional affiliations.

Printed on acid-free paper

This Springer imprint is published by the registered company Springer Nature Switzerland AG
The registered company address is: Gewerbestrasse 11, 6330 Cham, Switzerland

Contents

Chapter 1
Introduction

Armida de la Garza and Charles Travis

Abstract Nearly half a century after C.P. Snow's famous lecture on *The Two Cultures* (1959) which posited that a fracture in the intellectual life of the West existed between the science and the humanities, STEAM (Science, Technology, Engineering, Mathematics + Arts and Humanities) integrations and cross-pollinations are becoming more relevant than ever. Science can benefit from philosophical, ethical and aesthetic insights, in order to better deal with issues of uncertainty and contingency. Conversely, arts and humanities disciplines can be energized by scientific understandings of dynamic processes, technological innovations and the process of exploration and discovery. It is apt to recall that Leonardo Da Vinci's combined studies between 1490 and 1495 of art and science (particularly hydrology and the mathematics of perspective and proportion) contributed to his masterpieces *The Last Supper* (1498) and the *Mona Lisa* (1503). Equally, medical scientist James Lovelock's formulation of the *GAIA Hypothesis* (positing the earth as a self-regulating system) was informed in part by interactions with the writer William Golding, author of the novel *Lord of the Flies* (1954). Contemporarily, collaborative efforts between literary scholars and computational linguists have been able to trace the early onset of dementia in the works of authors Iris Murdoch, P.D. James and Agatha Christie. Indeed, as technology and the human species continue to symbiotically evolve, STEAM approaches will be crucial to facilitating acute and long-term insights into possible social and environmental interactions, impacts, benefits and consequences for our human condition. The present volume explores these exciting possibilities in detail, with contributions ranging from bacteria art, to the theoretical and practical benefits of dancing a PhD in renewable energy, to introductions to the emerging fields of heritage science, environmental and digital humanities, among others.

A. de la Garza (✉)
College of Arts, Celtic Studies and Social Sciences, University College Cork, Cork, Ireland
e-mail: adelagarza@ucc.ie

C. Travis
Department of History, University of Texas at Arlington, Arlington, TX, USA

Trinity Centre for Environmental Humanities, Trinity College, The University of Dublin, Dublin, Ireland
e-mail: charles.travis@uta.edu

© Springer International Publishing AG, part of Springer Nature 2019

A. de la Garza, C. Travis (eds.), *The STEAM Revolution*,
https://doi.org/10.1007/978-3-319-89818-6_1

Keywords STEAM · Interdisciplinarity · Transdisciplinarity · Art-science collaborations

Nearly half a century after C.P. Snow's famous lecture on *The Two Cultures* (1959) which posited that a fracture in the intellectual life of the West existed between the science and the humanities, STEAM (Science, Technology, Engineering, Arts and Humanities, and Mathematics)[1] integrations and cross-pollinations are becoming more relevant than ever.[2] Science can benefit from philosophical, ethical and aesthetic insights, in order to better deal with issues of uncertainty and contingency. Conversely, arts and humanities disciplines can be energized by scientific understandings of dynamic processes, technological innovations and the process of exploration and discovery. It is apt to recall that Leonardo Da Vinci's combined studies between 1490 and 1495 of art and science (particularly hydrology and the mathematics of perspective and proportion) contributed to his masterpieces *The Last Supper* (1498) and the *Mona Lisa* (1503).[3] Equally, medical scientist James Lovelock's formulation of the *GAIA Hypothesis* (positing the earth as a self-regulating system) was informed in part by interactions with the writer William Golding, author of the novel *Lord of the Flies* (1954).[4] Contemporarily, collaborative efforts between literary scholars and computational linguists have been able to trace the early onset of dementia in the works of authors Iris Murdoch, P.D. James and Agatha Christie.[5] Indeed, as technology and the human species continue to symbiotically evolve, STEAM approaches will be crucial to facilitating acute and long-term insights into possible social and environmental interactions, impacts, benefits and consequences for our human condition.

STEAM pathways will manifest differently, depending on interplays between public and developmental policies, specific regions' pedagogical and research organizations and the existence of entrepreneurial cultures or state led initiatives. Despite these differences, STEAM initiatives echo the legacy of liberal arts teaching and scholarship, the oldest academic tradition in the West. Liberal arts curricula comprise a tapestry woven from the study of literature, philosophy, history, art, drama, music and the social and physical sciences and attendant cognate disciplines. Contemporarily technical education and research rooted in the liberal arts

[1] Sometimes the 'M' is taken to mean 'Medicine', and sometimes the acronym is spelled STEAMM for the second 'M' to refer to medicine.

[2] Snow, Charles Percy. 2001. [1959]. *The Two Cultures*. London: Cambridge University Press.

[3] Nicholl, Charles. 2005. *Leonardo da Vinci: The Flights of the Mind*. Penguin.

[4] Lovelock, J.E. 1989. *The Ages of Gaia*. Oxford University Press, Oxford, UK.

[5] Le, X., Lancashire, I., Hirst, G., and Jokel, R. 2011. Longitudinal detection of dementia through lexical and syntactic changes in writing: a case study of three British novelists. *Lit Linguist Computing*. 26:4. pp. 435–461; Hirst G. and Wei Feng, V., 2012. Changes in style in authors with Alzheimer's disease. *English Studies*, 93:3. pp. 357–370; Hammond, A., Brooke, J. and Hirst, G., 2013. A tale of two cultures: Bringing literary analysis and computational linguistics together. In *Proceedings of the 2nd Workshop on Computational Literature for Literature* (CLFL'13), Atlanta.

(computing and engineering sciences, medical and film studies, informatics and data analysis) is a new thread running through this tapestry and one which has transformed the colour and texture of the liberal arts' fabric in significant and important ways. In many ways, this reconstituted tapestry has hastened the birth of the STEAM revolution.

With a resurgence of the liberal arts in Europe, and a crisis surrounding the liberal arts in the USA, STEAM approaches are becoming more relevant than ever. In Europe, where liberal arts scholarship and pedagogy were born, STEAM integrations are playing a role in re-cross-pollinating the arts, humanities and sciences. In America, where small colleges as well as faculties in larger universities espouse the liberal arts as a necessary foundation for functional literacy, leadership and citizenry in the democratic system, funding for the arts and humanities has been ostracized from the STEM disciplines.[6] This has been to the detriment of innovative STEAM teaching and research initiatives. The discussion of interdisciplinarity in higher education is now also growing in India and China. In the latter, the neologism 'kua4xue2ke1', translated literally as 'across subjects', was first introduced in 1986 but is only now starting to take off (Gombrich and Hogan 2017, p. 546).

Michael Lind, a professor at the Lyndon B. Johnson School of Public Affairs at the University of Texas, notes that 'in response to decades of slower- than-expected growth and heightened foreign competition, students deserted the humanities for more practical degrees like business'.[7] Lind states that in the first two decades of the twenty-first century, a bipartisan elite in the USA shared a consensus that national success depended not on the liberal arts but on proficiencies in science, technology, engineering and math. As a result, the humanities and specifically the study of history have been sacrificed on the altar of a new type of utilitarianism. Lind notes ruefully, 'the only academics who seem to find audiences among today's elite are economists and social scientists who claim to know how to boost gross domestic product or manipulate human behavior'.[8]

The mythical Cyclops in Homer's *Odyssey* (8 BC) reminds us of the drawbacks and dangers of such one-sided and short-sighted perspectives. In contrast, Ulysses managed over 12 ships and crews to navigate and overcome the obstacles he faced across the Ionian archipelago as he sailed home to Ithaca, to be reunited with Penelope after the fall of Troy. Similarly, STEAM integrations can be viewed as a choreography of ships from different disciplines, navigating the ontological and epistemological shoals which stand in the way of creating solutions to the tricky dilemmas posed by 'wicked problems' such as global climate change, public health

---

[6] Tubbs, N. 2015. *Philosophy and modern liberal arts education: freedom is to learn*. Basingstoke, Hampshire: Houndmills.

[7] Lind, M. 2017 Book Review Essay, "Why Arthur Schlesinger's 'Disunited States of America' Lives on", *The New York Times*, 2 Nov. 2017 <https://nyti.ms/2iVcD2D>.

[8] Ibid.

and safety, social-political-economic inequality and instability and the techno-cultural transformations and disruptions of the digital revolution, among others.[9]

The chapters opening the volume—the part entitled 'STEAM at work'—provide brief introductions to four emerging interdisciplinary fields in which STEAM collaborations are particularly fruitful. In Chap. 2, 'Spatial Humanities GIS: The City as a Literary, Historical and Cultural STEAM Lifeworld Laboratory', Charles Travis explains how geographic information systems (GIS) are being employed to discern relationships that make complex lifeworlds more immediately understandable by visually detecting spatial patterns that are normally hidden in texts or tables. Exploring methodologies in geo-data mining and the geo-semantic web, analysis and spatial visualization of cultural and historical data has opened new avenues of enquiry for the humanities and social sciences, which Travis illustrates with two case studies. The first one is a deep map of the life and works of poet Charles Bukowski, focusing on the places of his literary production by drawing on his poetry, prose and biography as data sources. The second discusses a deep mapping project of civilian deaths in Dublin that occurred during the 1916 Easter Rising, contextualized by its 2016 Commemoration reflected in digital and social media data. As a map does not just chart but 'unlocks and formulates meaning, [forming] bridges between here and there, between disparate ideas that we did not know were previously connected' (Cairo 2016, p. 263), the GIS-based deep mapping case studies discussed by Travis highlight how new meanings were generated and how space is understood as not simply the setting of action but a key product and determinant of change, in both the work of Bukowski and the interpretation of the history of the independence of Ireland.

In Chap. 3, 'STEAM Approaches to Climate Change: Extreme Weather and Social-Political Conflict', Francis Ludlow and Charles Travis argue that climate history and historical climatology can be clearly seen as an interdisciplinary STEAM frontier. The chapter discusses the potential and challenges of engaging in climate history, drawing upon case studies illustrative of the two main interrelated strands of work in the field. Adopting a teleological and didactic framing, the first strand is seen as a case of history informing science, taking the example of written evidence of the climatic impact of volcanism preserved within medieval *Irish Annals*. The second is viewed as a case of science informing history, taking the example of tree-ring-based evidence of past climatic extremes informing studies of the historical societal vulnerability to sudden environmental hazards. The role of climate as a trigger for human conflict is inevitably complex and mediated by the social and political background. Historians can supply nuanced and longer-term perspectives here. The degree to which climatic conditions may influence conflict is heavily contested, but remains under-researched with existing studies yielding contradictory results (Fjelde and von Uexkull 2012). Acute data shortages and the complexity of societal 'pathways' that may connect climate to conflict present major barriers to understanding (Scheffran et al. 2012). STEAM approaches which

---

[9]A wicked problem has been defined as one for which there can be no final solution, since any resolution generates further issues (Brown and Harris 2014, p. 3). Incomplete, contradictory and changing requirements that are often difficult to recognize make a solution impossible.

synthesize nomothetic and idiographic pathways can more comprehensively harvest, parse and analyse data, providing integrated multidimensional perspectives.

In Chap. 4, 'Film and the Medical Humanities: The "Romantic Science" of Neurocinema', Armida de la Garza and Germán Gil Curiel briefly trace the emergence of the medical humanities as an interdisciplinary field, focusing in particular on the connection between film studies and neuroscience. They first outline the pedagogical advantages that the use of films has brought to curriculum development in the health professions, and also the innovative applications of theories employed in the humanities to understand relations between doctors and patients. The chapter then presents a third, 'bio-cultural' paradigm in which films are employed for research on emotion, memory and in general cognitive processes, using fMRI and virtual neuron network simulations. They conclude with the application of a medical humanities framework for the analysis of the art film 'Requiem' by Alain Tanner (1998) which they argue can be read as humanizing the experience of schizophrenia and hallucination as opposed to construing it in terms of deficit and abnormality, which is usually the case. Their chapter closes with a plea for a post-disciplinary view of research in the arts and sciences that can focus on the potential for alleviating human pain, as opposed to focusing on the terms of disciplinary engagement.

The part closes with the report entitled 'Heritage Science' by Brendan Dooley and Armida de la Garza. Heritage Science is a growing interdisciplinary field that very aptly illustrates the benefits of STEAM in action. In brief, heritage science attempts to link specific competences in material science with the interpretative skills of the humanities and social sciences in order to understand and assist with the conservation of artefacts and sites that different cultures and societies value and regard as heritage, while also assisting with the curation of tangible artistic and archaeological capital of interest from economic and development perspectives. A report on a 2-day symposium on the subject held at University College Cork on 2–3 November 2017, the chapter illustrates what is at stake with heritage science through a variety of case studies, among them the following: the cross-referencing of the digital databases containing the archives of 1000 years of history of the city of Venice with a virtual reality projection to create 'the Venice Time Machine', which will show the way that news, money and commercial goods circulated in the city and point to migration and artistic/architectural patterns; the study of the various chemical substances that cause the characteristic smell of old books, and the recreation of this smell in the laboratory for its application to replicas in museums and libraries' special collections in order to add a sensory dimension to visitors' experience; and the use of nano-particles to repair works of art on torn paper, or paintings damaged by graffiti or other substances, or simply cleaning paintings on canvases that are hundreds of years old. Interestingly, heritage science is also the only STEAM interdiscipline in which the noun humanities is not linguistically modified by an adjective related to science, but the other way around.

The second part, Science and Art, presents three reflective essays, one from the perspective of a scientist engaging in artistic practice (dance) and the other two from artists engaging in scientific pursuits, one in biology and the other in ecology. It opens with the contribution by Aoife Long on her experience dancing her PhD in renewable energy for the 'Dance your PhD' competition, which she co-authored with Armida de la Garza. In the chapter, Long explains how this facilitated not only the transmission and dissemination of her research in a highly engaging way and allowing her to reach a wider audience but how it also led to the construction of embodied knowledge, an important development in the learning of science, arrived at by means of the arts via the construction of conceptual metaphors.

Chapter 7 is written by Anna Dumitriu, who could be described as an artist that works through the medium of bacteria. Her chapter recounts numerous projects in which the work has been, as she puts it, both about creating understanding of a scientific subject for the general public and also about feeding back into patient management strategies some of what was learned from actually working with the general public. For instance, she used a seaweed-based growth medium that makes bacteria grow in a different colour in accordance to their resistance of specific antibiotics and which made the superbug form of *Staphylococcus aureus* grow into what the doctors in her team described as 'a lovely denim blue'. She then inserted this into cotton to produce patches to make a quilt that is a work of art in its own right, for an exhibition about bacteria. Further, while Dumitriu describes her aim as 'to communicate the impact of new technologies in microbiology and how they will improve understanding, diagnosis, treatment and control of infections as well as to think critically about them and bring an artistic voice into the lab' (p. X), she contends that 'the choices that scientists make, the passionate ones at least, can be very much aesthetically driven' (p. X).

Chapter 8 is authored by artist/educator Anita McKeown. Drawing from the principles of permaculture, a system of agricultural and social design focused around simulating or imitating the patterns and features observed in natural ecosystems and now expanded to stand also for 'permanent culture', McKeown explains how STEAM informs her aesthetic and pedagogical practice, which is based in situated art. Using the creation of the Beautiful Mitten School curriculum as a case study, McKeown explains how a STEAM approach goes well beyond the mere use of art as a tool or aid in the teaching of science, becoming instead a method that integrates learning into a holistic experience, addressing the complex problems of the twenty-first century, and rooted in specific communities and ecologies.

While the projects described in this part are themselves very different, what the chapters have in common is a testimony on the effectiveness of the interdisciplinary collaboration for teaching and learning and to elicit the active engagement of a variety of stakeholders. They also highlight the personal and professional development and growth that the three authors have experienced as a result.

Part III is on facilitating and supporting STEAM collaborations, and it comprises two pieces with a practical focus and one that focuses instead on a conceptual framework that sets out the social need, even urgency, to facilitate and support this collaboration. Chapter 9 by Annmarie Ryan and her team describes the methodology followed by their Health Research Futures Lab in Limerick to set up interdisciplinary research teams and guide them through the process of developing a research project proposal to the stage of grant application and actually seeing a project resulting from the experience implemented. The key to their method is a 4-day retreat based on a model developed at the University of Limerick by an interdisciplinary group from the faculties of Science and Engineering (STEM), the Kemmy Business School and the Irish World Academy of Music and Dance (AHSS). Their chapter provides a thorough assessment of the challenges and benefits inherent in this approach and is detailed enough that it can provide a guide for institutions and/or academics, artists and practitioners interested in cultivating STEAM interdisciplinary research.

Chapter 10 by Armida de la Garza is entitled ‘Internationalisation in Higher Education as a Catalyst to STEAM’. The chapter argues that the internationalization of the curriculum agenda in higher education partly inherently overlaps with that of STEAM cultivation, and it highlights two practical ways in which curricula can be modified to promote the latter while advancing the former for a more inclusive student experience, enhancing employability skills and promoting the interdisciplinary outlook to the most pressing wicked problems that societies so badly need to address today.

The last chapter in this part entitled ‘What Can We Learn About STEAM from Bridges?’ by David Blockley traces the process whereby architecture, engineering and the arts and crafts became separated from each other, since the four of them were actually part of the same knowledge base before the Renaissance and were indeed in a sense also intertwined with science, as Blockley shows. His examination continues to the present day, when he contends engineering has come to be regarded as applied science—technology. Engineering was thus downgraded in popular culture to craft, because it is not understood as craft in its creative sense ‘but more akin to vocational technique and manual skills devoid of creativity. This has been and still is a major category mistake’ (p. X).

In a sense, bridges themselves are a metaphor of STEAM at work, not only because they connect ‘territories’ and provide a means for crossing, but also because they are works of engineering, involving science, and, in some cases, also real works of art. As an essential part of public infrastructure, the metaphor highlights the need to build them everywhere in the ‘public infrastructure’ of academia too. Moreover, as Blockley clearly explains, managing tension is in fact the key central aspect to a bridge standing, again exactly the case when it comes down to academic disciplines. Blockley’s chapter illustrates this by attempting to, as it were, become a bridge itself, using the concept of energy. ‘In all systems, there is a potential that drives flow against an opposition. In this chapter I describe these processes for bridges. I […] show how forces flow through a structure. [This will] illustrate the art of STEAM’ (p. X).

Part IV ‘STEAM Teams’ provides examples of research in action. Like the part above, it comprises two practical and one conceptual essays. In Chap. 12, Vikram Pakrashi and his team provide an example of multidisciplinary research at work comprising the social sciences (psychology), engineering and medicine. Their chapter describes a collaborative project designed to measure the impact of cyclists’ perceptions of risk on their heart-beat data produced by the cyclists’ wearable technology. In addition to the actual experiment design and implementation, the chapter also describes the challenges the team faced to be able to work together given the diversity of their backgrounds and some reflexion on what in their view was gained by the exercise.

Chapter 13 by Donal Healion et al. carefully documents the 'Practice-based Experiential Learning Analytics Research and Support (PELARS)' project. This was a 3-year EU-funded project comprising 12 partners drawn from academic, corporate and non-profit sectors, located across 9 countries. The premise of the project is that the physical form of furniture elements within a learning environment has a bearing on group formation and dynamics (Healion et al. 2017). Because all collaboration is a product of social interaction, by providing suitably designed furniture, students can work physically closer together and move more easily, which facilitates greater interaction and thus collaboration both among group members and between groups. The PELARS project thus involved the design of a learning environment featuring modular and mobile educational furniture in which technology to facilitate learning was embedded. The furniture is designed to foster communication and interaction, such as peer to peer and student to teacher, during collaborative hands-on activities, facilitating the documentation of ideas and allowing for ease of movement and postural changes. The chapter provides guidance and advice to prospective partnerships intending to carry out transdisciplinary STEAM-related research.

In general, design has been acknowledged as inherently interdisciplinary, straddling craft and science as well as the humanities, the social sciences and engineering in that designers always have to take into consideration various concerns: aesthetic, ergonomic, safety, accessibility, marketability, profitability and sustainability, among others (Boradkar 2017, pp. 456–457). For this reason, the kind of thinking required for design can integrate multiple points of view and has been hailed as a means to promote interdisciplinarity, in particular STEAM: 'designers need a unique set of tools [...] brainstorming, mind mapping, visualisation, prototyping, storyboarding, scenario development' (Boradkar 2017, p. 465) and so on. The usefulness of design thinking for interdisciplinary endeavours is also evident in Chap. 9, as a key component of the Future Health Research Lab is precisely design thinking.

The last chapter in this part, Chap. 14, by Francesco Dattilio and Valeria De Pascale, briefly traces the history of the concept of the Anthropocene, explaining why only a STEAM theoretical framework can be useful to understand what is at stake in both the philosophical concept itself and the historical period it refers to.

Part V is on policy, for 'if knowledge is to be genuinely interdisciplinary, it needs to do more than simply reach across campus [...] Our academic research portfolio must include an account of how to effectively integrate knowledge with the decision-making context faced by governments, businesspeople and citizens' (Frodeman and Mitcham 2005, p. 513). Chapter 15 by Marie Clarke contends that for STEAM to advance beyond individual initiatives or transdisciplinary/interdisciplinary debates, it has to become part of the 'policy window' where policy issues move onto the government agenda and result in decision and action (Kingdon 1995). Using Kingdon's Multiple Streams Framework (MSF), Clarke explores the policy emphasis on STEM in the light of the 2007 financial crisis, which resulted in 9 years of austerity measures in Ireland (and elsewhere in Europe) and assesses the feasibility of introducing the arts, humanities and social sciences into the mix.

Chapter 16 by Leonard Hobbs argues in favour of the promotion of the concept of STEAM compared to just STEM. It reflects on the lessons from history by looking at some of the world's greatest innovators who were practitioners of STEAM even though it was not called thus then, and whose many breakthroughs were enabled by a combination of the creative and innovative processes. It goes on to describe how diversity in the workplace, which is an accepted catalyst for business improvements, can be optimized by the application of a STEAM agenda by enterprises.

Taken together, the chapters in this part thus bring governments, businesses and enterprises into the picture.

Finally, Chap. 17 by Robert Arnold closes the volume with a reflective piece that proposes a critique of STEAM. Arnold argues that art and science and technology and perhaps to a somewhat lesser degree engineering and math have always been important, vital, aspects of art on many different levels. They have always been interconnected. By the same token, the areas that make up STEM have never lacked creativity. Indeed Arnold states that art cannot claim a monopoly over creativity as a force that drives all human activity. The drive to discover, to know, to express and to solve problems, technical or otherwise, is always a creative act. However, Arnold's chapter is underscored by this question: simply how does the potential of art to work against the grain of dominant ideology, to distance us from the mechanisms that construct our prevailing understanding of reality, or even to offer a moment of perception that is not devoted to practical utilitarian pursuits, fit within the STEAM concept of serving primarily economic goals?

The chapters of this edited volume are rooted in a variety of disciplinary perspectives. Subsequently, we as editors have decided to eschew uniformity in referencing and annotation styles. Therefore, to highlight the variety, differences and contrasting theories, methods and modes of scholarship and research coalescing under the STEAM umbrella, each chapter's referencing and annotation styles will conform to the disciplinary convention practiced by their respective authors. In this way, the volume is both broad in scope and diverse as regards the nature and themes of the contributions. It offers theoretical frameworks, concepts and theories, as well as practical advice and guidance on the cultivation of STEAM. This will make it relevant for academics, artists, practitioners, funding bodies and businesses alike. All welcome to the STEAM revolution!

References

Altbach PG, Yudkevich M (2017) International faculty in 21st century universities: themes and variations. In: Yudkevich M, Altbach PG, Rumbley LE (eds) International faculty in higher education: comparative perspectives on recruitment, integration and impact. Routledge, London, pp 1–14

Boradkar P (2017) Taming wickedness by interdisciplinary design. In: Frodeman R (ed) The Oxford handbook of interdisciplinarity. Oxford University Press, Oxford, pp 456–467

Brown VA, Harris JA (2014) The human capacity for transformational change: harnessing the collective mind. Routledge, London/New York

Cairo A (2016) The truthful art: data, charts and maps for communication. New Riders, Berkeley
Frodeman R, Mitcham C (2005) New directions in interdisciplinarity: broad, deep and critical. Bull Sci Technol Soc 27(6): 506–514
Gombrich C, Hogan M (2017) Interdisciplinarity and the student voice. In: Frodeman R (ed) The Oxford handbook of interdisciplinarity. Oxford University Press, Oxford, pp 544–557
Kingdon J (1995) Agendas, alternatives and public policies. Longman, New York
Leask B (2015) Internationalising the curriculum. Routledge, London
Pacheco RCS, Manhaes M, Maldonado MU (2017) Innovation, Interdisciplinarity and Creative Destruction. In: Frodeman R (ed) The Oxford handbook of interdisciplinarity, 2nd edn. Oxford University Press, Oxford, pp 303–318

Part I
STEAM at Work: Emerging Interdisciplinary Fields in the Sciences, Arts and Humanities

Chapter 2
Spatial Humanities GIS: The City As a Literary, Historical, and Cultural STEAM Lifeworld Laboratory

Charles Travis

> *Behavior in space and time [is like] the surface movements of icebergs, whose depths we can sense only vaguely. Anne Buttimer, Grasping the Dynamism of Lifeworld (1976)*

Abstract Cities envisioned through the phenomenological lenses of literature and history cannot necessarily be parsed by algorithms, sensors and Smart City techno-rhetoric, but may suggest STEAM framed avenues of engagement between literary, historical and cultural scholars and computer scientists and urban engineers. As a STEAM cognate discipline, the Spatial Humanities is a new interdisciplinary field which finds scholars in literature, history, philosophy, linguistics, film, culture and media studies engaged with colleagues from scientific and quantitatively-oriented disciplines to explore questions on geographical and conceptual space. The field exploits various geospatial and digital data technologies, guided in their use by the history and philosophy of spatial thought, literary criticism, linguistics, digital media and geographic information science (GIS). By focusing questions concerning place and space through the lens of the humanities, the field has been exploring methodologies in geo-data mining, the geo-semantic Web, and the visualization, analysis and spatial applications of cultural and historical data, among other topics. The agenda of the Spatial Humanities includes the pursuit of theory, methods, case studies, experiments, applied technology, broad narratives, and more persuasive strategies. Its aims are commensurate with STEAM initiatives in its goal to bridge research fields in arts, humanities and (natural, life and social) sciences.

Keywords GIScience · Smart cities · Literature · Public history · Social media

C. Travis (✉)
Department of History, University of Texas at Arlington, Arlington, TX, USA

Trinity Centre for Environmental Humanities, Trinity College, The University of Dublin, Dublin, Ireland
e-mail: charles.travis@uta.edu

© Springer International Publishing AG, part of Springer Nature 2019
A. de la Garza, C. Travis (eds.), *The STEAM Revolution*,
https://doi.org/10.1007/978-3-319-89818-6_2

Introduction

Cities envisioned through the phenomenological lenses of literature and history cannot necessarily be parsed by algorithms, sensors, and smart city techno-rhetoric, but may suggest STEAM-framed avenues of engagement between literary, historical, and cultural scholars and computer scientists and urban engineers. As a STEAM cognate discipline, the Spatial Humanities is a new interdisciplinary field which finds scholars in literature, history, philosophy, linguistics, film, culture, and media studies engaged with colleagues from scientific and quantitatively oriented disciplines to explore questions on geographical and conceptual space. The field exploits various geospatial and digital data technologies, guided in their use by the history and philosophy of spatial thought, literary criticism, linguistics, digital media, and geographic information science (GIS). By focusing questions concerning place and space through the lens of the humanities, the field has been exploring methodologies in geo-data mining, the geo-semantic Web, and the visualization, analysis, and spatial applications of cultural and historical data, among other topics. The agenda of the Spatial Humanities includes the pursuit of theory, methods, case studies, experiments, applied technology, broad narratives, and more persuasive strategies. Its aims are commensurate with STEAM initiatives in its goal to bridge research fields in arts, humanities, and (natural, life, and social) sciences.

This chapter will discuss Spatial Humanities GIS deep mapping methodologies, as they relate to smart cities, "sense of place," and distant and close reading techniques. These theoretical approaches with be applied to two STEAM case sketches. The first sketch will discuss deep mapping the "Dirty Realism" of Los Angeles "Skid Row" poet, Charles Bukowski, focusing on the places of his literary production, by drawing on his poetry, prose, and biography as data sources. The second sketch will discuss STEAM approaches to a public history Spatial Humanities deep mapping of civilian deaths in Dublin that occurred during the 1916 Easter Rising, contextualized by its 2016 Commemoration reflected in digital and social media data. Both sketches will engage Spatial Humanities *distant* and *close* reading techniques.

Deep Mapping, Spatial Humanities GIS, and the Smart City

GIS deep mapping is one emerging technique and holds considerable potential for integrating both Spatial Humanities and STEAM research into smart city computing and engineering approaches. The Native American writer William Least Heat-Moon (a.k.a. William Lewis Trogdon) first employed deep mapping as a discursive, stratigraphic literary method to explore the "sense of place" of a single county on the plains of Kansas in *PrairyErth: A Deep Map* (1991). Deep mapping has been employed as a heuristic by Spatial Humanities GIS practitioners to design mixed method (qualitative and quantitative) data structures and models to map the dynamic stratigraphies of period and place. Deep mapping weaves oral testimony, anthology, memoir, and biography into a "vertical form" of travel writing to reveal the grain

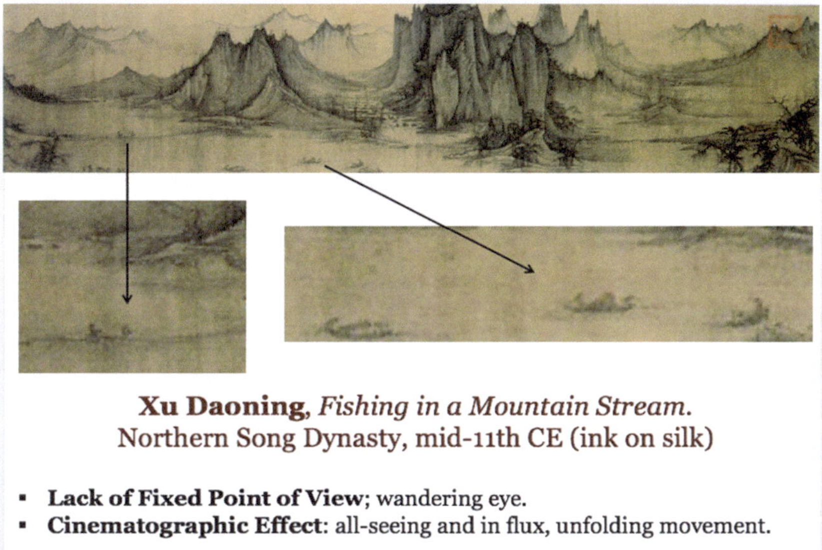

Fig. 2.1 Mid-eleventh-century Chinese landscape painting and the "Drifting Cinematographic Eye"

and patina of place by intersecting the historical with the present, the political with the poetic, and the discursive with the sensual.[1] Least Heat-Moon can be considered a topographical travel writer, and his works include *Blue Highways: A Journey Into America* (1982) and *River Horse: The Logbook of a Boat Across America* (1999). He also translated and edited *An Osage Journey to Europe 1827–1813* (2013) symbolizing the confluence of two aesthetic and intellectual traditions. The first concerns occidental perspectives on space and place described previously in this chapter, and the second is reflecting subtly the imprint of the oriental, as his Native American ancestral footprints arguably include forbearers who migrated a millennium ago across the now submerged Siberian land bridge. In this sense, his works impart cinematographic, oriental, and phenomenological perspectives which can inform the development of GIS deep mapping sensibilities. The cinematographic eye offers a "way of seeing," similar to the ocular technologies, techniques, and discursive tropes such as "landscape" that have preoccupied the practices of geography, cartography, and GIScience (Fig. 2.1).

According to geographers Marcus Doel and David Clarke (2007), cinematic perspectives possess the "ability to abstract, manipulate and reengineer the spatial and temporal registration of events."[2] This phenomenological ability, according to Siegfried Kracauer, allows the cinematographic eye to "drift"

[1] Pearson, M. and Shanks, M. 2001. *Theatre/Archaeology - the (re)articulation of fragments of the past as real-time event*. London, Routledge.

[2] Doel, M.A. and Clarke, D.B., 2007. Afterimages. *Environment and Planning D: Society and Space*, 25:5. Pp. 890–910.

> [...] towards and into the objects -so much like the legendary Chinese Painter who, longing for the peace of landscape he had created, moved into it, walked towards the faraway mountains suggested by his brush strokes, and disappeared into them never to be seen again.[3]

Conceiving and incorporating phenomenology and cinematographic and oriental aesthetics into geospatial technology models to represent past and present "senses of place" raises significant representational challenges. However, currently GIScience urban "Platial" models are beginning to integrate social media data with Big Data streams to facilitate a significant conceptual shift from the classical "layer-cake view of the world" to a digital "networked cupcakes view of the world."[4] Relevant to the evolution to Spatial Humanities GIS deep mapping practices, such "Platial" models illuminate Jaime Lerner's theory of "urban acupuncture" in which a city is viewed as a living organism possessing specific "neural" target points that can be targeted and engaged to re-energize its corpus.[5] Connecting the "dots" of these target points reveals what Seamus Deane, parsing the work of Walter Benjamin, defines as a "constellation":

> [...] a previously unrecognized structure or network of relations that was always there, like the unconscious, and appears to us, like it, in articulated images, laden with the weight of the past and yet haloed in the light of discovery and recognition.[6]

Such a constellation was revealed by a Spatial Humanities GIS social media survey of a recent Bloomsday celebration in Dublin. This deep mapping experiment employed James Joyce's novel *Ulysses* (1922) as a discursive structure to plot a narrative set by Joyce in 1904 upon the constellated "senses of place" generated by Joycean pilgrims in the Irish capital on 16 June 2014 (Fig. 2.2).[7]

In the zeitgeist of the early twenty-first century, such Spatial Humanities approaches and deep mapping experiments are useful in offering a different perspective on the relationship between cities, human agency, and the "wicked problems" of the twenty-first century such as global warming and political polarization. Indeed, the majority of the world's population will be residing in urban areas by 2050; as such, cities have become primary sites of social experimentation and problem-solving, and the smart city heuristic has been promoted as "a somewhat nebulous idea that seeks to apply massive amounts of digital data collected about society as a means to rationalize the planning and management of cities."[8] The smart city, being

[3] Kracauer, S., 1960. *Theory of film: The redemption of physical reality*. Princeton University Press. 165

[4] Roche, S. 2015. 'Geographic information science II: Less space, more places in smart cities' *Progress in Human Geography*. 1–10. Accessed 1 June 2015 at: <http://phg.sagepub.com/content/early/2015/05/19/0309132515586296.full.pdf+html>.

[5] Lerner. J. 2014. *Urban Acupuncture*. Washington, DC: Island Press.

[6] Deane, S. (2007). Walter Benjamin: The Construction of Hell. *Field Day Review*. 3:2–27. Pg. 10.

[7] Travis. C. 2015. Visual Geo-Literary and Historical Analysis, *Tweetflickrtubing*, and James Joyce's *Ulysses* (1922). *Annals of the Association of American Geographers*<https://doi.org/10.1080/00045608.2015.1054252

[8] Shelton, T., Zook, M., & Wiig, A. (2014). The 'actually existing smart city'. *Cambridge Journal of Regions, Economy and Society*, rsu026.

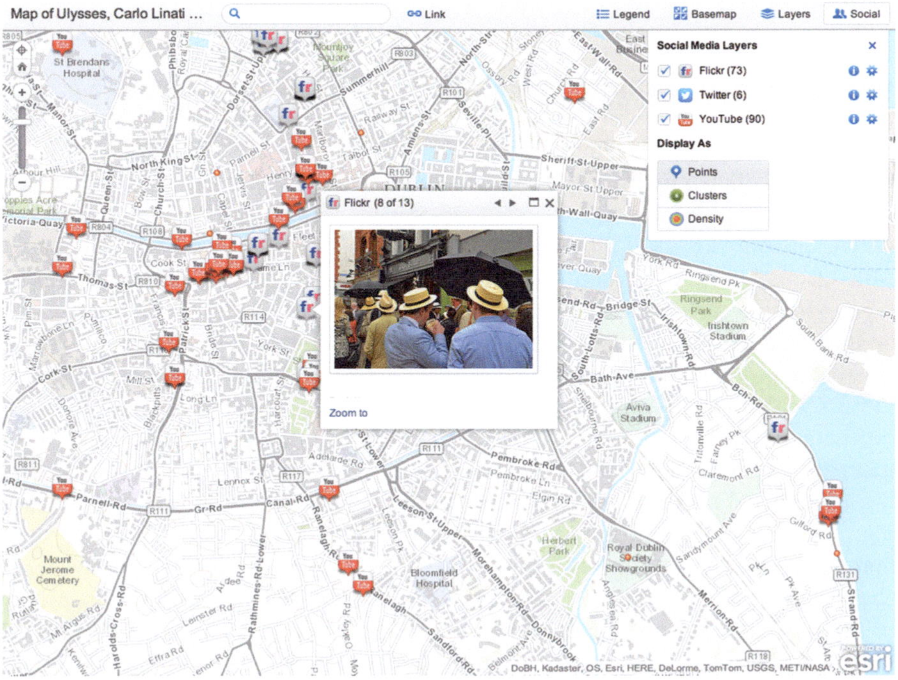

Fig. 2.2 Bloomsday "sense of place," Dublin 2014 (Charles Travis)

the latest iteration of an "urban science" discourse (similarly promulgated by the "quantitative revolution" in geography and planning in the 1950s), frames the city in a cybernetic webwork of information communication technology (ICT) global positioning satellites (GPS) and GIS. By harvesting digitized social and environmental data and the coding and algorithmic programing of various activities, urbanity and the human condition are conflated as a "complex network of interconnected systems," to be designed and regulated for the common social and environmental good.[9]

Anticipating these emerging forms of technically mediated urbanity, Henri Lefebvre warned of the *cyberanthrope* whom "disqualifies humanism in thinking and action" by purging the "illusions of subjectivity: creativity, happiness, passion" and severely treating "the dramatic, the historic, the dialectic, the imaginary, the possible-impossible" by living "in close proximity with the machine."[10] Subsequently, transdisciplinary dialogue, reflecting STEAM approaches, between human geographers and scholars with spatial and urban interests, has made gains in theorizing the human condition and its wider relationship between GIS, ICT, digital data, transportation and mobility, and smart cities.

---

[9] IBM Smarter Cities. 2012. *Smarter, More Competitive Cities Forward-thinking Cities Are Investing in Insight Today*. Somers, NY: IBM.

[10] Lefebvre, H. 1971 *Vers le cybernanthrope: contre les technocrats*. Denoël: Paris. Pp. 194–198.

GIS and Sense of Place

Donald J. Janelle (2001) states that underlying complexities in the human organization of urban space present methodological problems for GIS in linking empirical research questions with alternative theoretical frameworks.[11] However, GIS and humanities collaborations have been facilitated by the digital revolution of the last quarter century. In the 1950s, Claude Lévi-Strauss outlined the "three humanisms" of Western history; in the twenty-first century, a "fourth humanism," coined by Milad Doueihi (2011) as "digital humanism," describes a "type of society," multiple types of media and texts (books, maps, multimedia, augmented reality), that cannot be fixed in space or stabilized over time."[12] "Digital humanism" coincides with the dissolution of epistemological boundaries between science and technology studies, the arts, and humanities. Additionally, the rise of social media has contributed to the evolution of digital humanities theory and practice. The digitization of historical, literary, and artistic texts and emergence of online research methods and teaching have dovetailed with humanities computing quantification projects and digital parsing, analysis, and visualization applications. Human geographers have long held that literary texts are the product of human perception and consider discursive tools such as imagery, narrative, and setting as means to access the subjective dimensions of a particular location's "sense of place."[13] In *Humanistic Geography* (1976) Yi-Fu Tuan considered five elements constituting a "sense of place": the nature of geographical knowledge, the role of territory in human behavior, the creation of place identities, the role of knowledge as an influence on livelihood, and the influence of religion on human activity. Ideology, culture, affect, and as Tuan observed "sense of time" affect a location's sense of place.[14] In contrast, the Irish poet and Nobel laureate Seamus Heaney defines a "sense of place" from a purely literary perspective:

> [...] it is this feeling, assenting, equable marriage between the geographical country and the country of the mind, whether that country of the mind takes its tone unconsciously from a shared oral inherited culture, or from a consciously savoured literary culture, or from both, it is this marriage that constitutes the sense of place in its richest possible manifestation.[15]

However, such nuanced, affective, and subjective perspectives on the human experience of place have been elided in most GIS research and study. Daniel Sui and Michael Goodchild note: "until recently, place has been off the intellectual radar screen of GIScientists, many of whom appear to use the two terms place and space

---

[11] Janelle, D.G. 2001. Time-space. In Geography. *In*: N.J. Smelser and P.B. Baltes, eds. *International Encyclopedia of the Social and Behavioral Sciences*. Amsterdam: Pergamon-Elsevier Science, 15746–15749.

[12] Doueihi, M. 2011. *Pour un humanisme nume'rique*. Paris: Seuil.

[13] Travis. C. 2006. *Lifeworlds: Literary Geographies in 1930s Ireland.* PhD thesis, Trinity College Dublin; Tuan, Y-F. 1977. *Space and Place: The Perspective of Experience.* London: Arnold.

[14] Tuan, Y-F. 1976. 'Humanistic Geography.' *Annals of the Association of American Geographers*, 66:2. Pp. 186; 266–276.

[15] Heaney.S. 1980. *Preoccupations, Selected Prose 1968–1978*. N.Y. Farrar, Straus, Giroux. pg. 132.

somewhat interchangeably."[16] In contrast, the objective of GIS deep mapping to quote Trevor M. Harris "is to shift from a view of humans as entities or data points to an examination of behavior, the material and imaginary worlds, and the relationships that compose notions of a nuanced, non-reductionist, deeply contingent, and scaled conception(s) of place."[17] Recently, commensurate GIS applications have engaged the study of urbanity as perceived in literary, historical, and cinematic texts, in addition to the perception and the remediation of place diffracted by social media platforms and mobile computing devices.[18]

As Shannon Mattern (2016) observes, the creation of texts and urbanities has been intertwined for at least a millennium:

> Some of the first writing surfaces, clay and stone, were the same materials used to construct ancient city walls and buildings, whose facades also frequently served as substrates for written texts. The formal properties of those scripts—the shapes they took on their clay or, eventually, parchment and paper foundations—were also in some cases reflected in urban form: how the city molded itself from the materials of the landscape. And those written documents have always been central to our cities' operation: their trade, accountancy, governance, and culture.[19]

One of the key concerns underpinning Spatial Humanities, and deep mapping approaches, is how geospatial technologies can be used to mine, manage, manipulate, chart, visualize, and analyze the subjective geographies embedded within such documents and literary, historical, and cultural texts.[20] Paul Ricoeur observes that literature, born from the life of writers, provides "an unstable mixture of fabulation and actual experience" and thus "an immense laboratory for thought experiments."[21] In *Berlin Chronicle* (1927–1934) Walter Benjamin's study of the city, the literary critic

> [...] played with the idea of setting out the sphere of life-bios-graphically on a map. First I envisaged an ordinary map, but now I would incline to a general staff's map of a city center, if such a thing existed.[22]

---

[16] Sui, D. and Goodchild, M. 2011. 'The convergence of GIS and social media: challenges for GIScience' *International Journal of Geographical Information Science*. 25(1): 1737–1748.

[17] Harris, T.M. 2015. Deep Geography-Deep Mapping: Spatial Storytelling and a Sense of Place in *Deep Maps and Spatial Narratives*. Ed. David J. Bodenhamer, John Corrigan, and Trevor M. Harris. Bloomington & Indianapolis: Indiana University Press. Pg. 42.

[18] Travis, C., 2015. *Abstract machine: humanities GIS*. Esri Press.; Cooper, D., Donaldson, C. and Murrieta-Flores, P. eds., 2016. *Literary mapping in the digital age*. Routledge.; Stadler, J., Mitchell, P. and Carleton, S., 2015. *Imagined landscapes: geovisualizing Australian spatial narratives*. Indiana University Press.

[19] Mattern,S. 2016. "Of Mud, Media, and the Metropolis: Aggregating Histories of Writing and Urbanization," *Cultural Politics* 12:3 (November 2016): 310–31. Pg. 310.

[20] Gregory, I., Bushell, S. and Cooper, D., 2013. *Mapping the Lakes: Towards a Literary GIS*.

[21] Ricoeur, P. 1992. *Oneself as Another*, trans. Kathleen Blarney. Chicago, IL: University of Chicago Press. pg.159.

[22] Benjamin, W. 1999. 'A Berlin Chronicle.' *Walter Benjamin: Selected Writings*, Volume 2: 1927–1934. Ed. Michael Jennings, Howard Eiland and Gary Smith. Trans. Edmund Jephcott. Cambridge. Harvard UP. Pp. 595–637.

Dan McQuillan (2017) echoes this perspective, stating that smart cities are "more clearly revealed by the idea of the map as narrative, rather than as a spatial construction; that is, the map is something constructed out of movements and histories rather than something that precedes them."[23] Doueihi's (2013) perception of digital humanism asks "what is the situation with the anthropology of this new inhabited earth, these new digital territories that are flexible, fluid and constantly moving? How should we think about them, analyse them, especially since geolocation and smart cities cannot be dissociated from our daily lives?"[24]

Such a perspective, anticipated by Henri Lefebvre in *The Production of Space* (1991), considered the "hyper-complexity of space" in which "fixed points, movements and flows and waves—some interpenetrating, others in conflict," coalesce to compose the human fabric of a city.[25] Stephen Roche's (2015) smart city schema sharpens Lefebvre's perspective in four ways: firstly as an intelligent city (social infrastructure); secondly, as a digital city (informational infrastructure); thirdly, as an open city (open governance); and fourthly, as a live city (a continuously adaptive urban living fabric). Consequently, smart cities can be perceived as a human web of *places* rather than an abstract network of *spaces*. Urban areas can now be surveyed and mapped textually, visually, and aurally through geospatial lenses to identify and analyze "the digital (spatial) activity generated by social media users."[26] While GIS is a digital tool, the humanities are still overwhelmingly conceived and practiced as analogue disciplines and have yet to reap fully from the benefits of the twenty-first digital revolution. Digitization, the first steps to new procedures of inquiry in the "Age of Big Data," has enabled scholars to collate and parse geo-data sets and ask questions that would have been considered impossible at more than a speculative level only a decade ago. Indeed, analogue archives hold seas of human knowledge, which have yet to be digitally trawled to their true depths. Locked in parchment, paper, pieces of photography, cartographical documents, paintings, and microfilm, such academic artifacts have become attribute sources for Spatial Humanities GIS deep mapping expeditions.

GIS and Distant and Close Readings

The deep mapping of Bukowski's Los Angeles and Dublin 1916/2016 by GIS in this chapter is guided by contrapuntal *distant* and *close* readings of contrasting literary and historical perceptions of urban space. Historical, literary, and biographical texts

[23] McQuillan, D. 2017. Countermapping the SmartCity, *Navigations* (Spring).

[24] Doueihi, M. 2013. 'About Digital Humanism.' *Ideas*. Accessed August, 15 2015. <http://www.inaglobal.fr/en/ideas/article/about-digital-humanism>.

[25] Henri Lefebvre in *The production of space* (1991) (2007, 88).

[26] Roche, S. 2015. 'Geographic information science II: Less space, more places in smart cities' *Progress in Human Geography*. 1–10. Pg. 6. Accessed 1 June 2015. <http://phg.sagepub.com/content/early/2015/05/19/0309132515586296.full.pdf+html>.

can be visually juxtaposed, parsed, plotted, and contextualized by GIS with an analysis of the historical and cultural landscapes composing the city. *Distant reading*, coined by literary scholar Franco Moretti, approaches literature "not by studying particular texts, but by aggregating and analyzing massive amounts of data," and employing "statistical, quantitative methods to 'read' large volumes of text at a distance, using 'graphs, maps, and trees' as forms of abstract representation that enable the study of patterns over time."[27] In contrast, *close reading*, a fruit of New Criticism theory, undertakes careful, sustained, deconstructive interpretations on brief passages of text. Translating these two critical types of reading techniques into GIS deep mapping creates differing scales from which to map the works and life spheres of writers, their personas, and characters. By employing Benjamin and Lefebvre's cartographical aspirations and spatial sensibilities, the works of Bukowski and the 1916 Easter Rising/2016 Commemoration digital narratives will plot deep mappings of Los Angeles and Dublin, respectively, to consider how the milieus and paths of fictive and active *lifeworlds* intersect and interpenetrate to compose the human fabric of the city.

Los Angeles, United States of America: Charles Bukowski's "Dirty Realism"

This urban deep mapping sketch considers a similar urban biographical study of the writer Charles Bukowski's (1920–1994) poetic production. Los Angeles has been acknowledged as "Buk" territory, and he distilled the personas and places in his poetry and prose from his various guises as a skid row bum, wage slave in dead-end jobs, post-office employee, and charity case suffering from liver malfunction.[28] Born in Andernach, Germany, Charles "Hank" Bukowski and his family moved to South-Central Los Angeles when he was 3. Growing up, Bukowski suffered an extreme case of acne and was bullied by his father and his classmates. An old brownstone near 21st Street and La Brea Avenue housed the local library of his childhood and provided him a safe haven in his early teens where he discovered and learned to admire the writers Sinclair Lewis, D. H. Lawrence, and Ernest Hemingway. After Bukowski's first taste of alcohol as a teenager (provided by his friend William "Baldy" Mullinax who appears in the semiautobiographical novel *Ham and Rye* (1982) as "Eli LaCrosse"), Bukowski realized that drinking was "going to help me for a very long time."[29] He graduated from Los Angeles High School and attended Los Angeles City College for 2 years, taking courses in art, journalism, and litera-

[27] Moretti. F. 2000. Conjectures on World Literature. *New Left Review* 1: 54–68; Franco Moretti. 2013. Operationalizing: Or, the Function of Measurement in Modern Literary Theory.' *Pamphlets of the Stanford Literary Lab* 6:1–13 <http://litlab.stanford.edu/LiteraryLabPamphlet6.pdf.>.

[28] Madigan, A. 1996. What Fame Is: Bukowski's Exploration of Self. *Journal of American Studies, 30*(3), 447–461.

[29] Bukowski, C., 2009. *The Roominghouse Madrigals: Early Selected Poems 1946–1966*. Harper Collins.

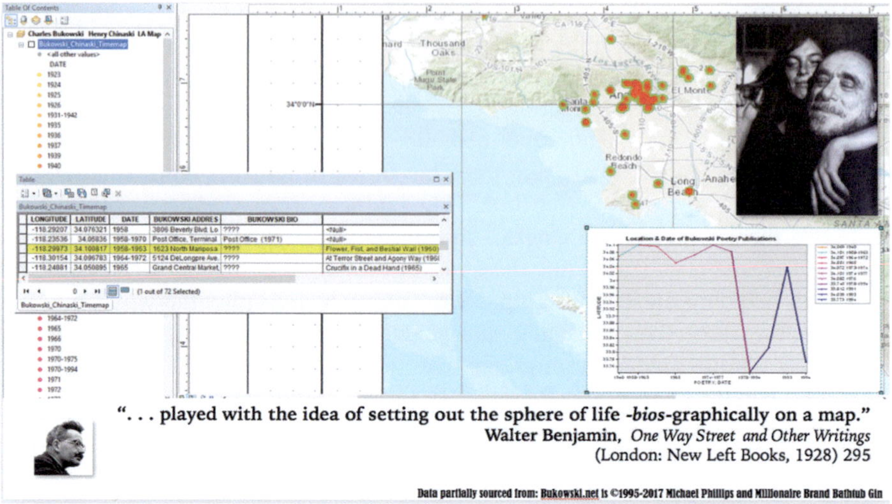

Fig. 2.3 Charles Bukowski's "Skid Row Latitude" of 34°0′0′ (Charles Travis)

ture, before dropping out at the break of the Second World War. In the 1940s and 1950s, he traveled around the United States working as an itinerant laborer, depicted in his novel *Factotum* (1975) before returning to California. Bukowski's restlessness is captured in his poem *Consummation of Grief*:

I was born to hustle
Roses down the avenue
Of the dead.[30]

In 1967, Bukowski began writing for the alternative Los Angeles paper *Open City*. Bukowski's poetry, autobiographical stories, essays, and other musings were published alongside "coverage of student unrest, the New Left, black power, civic and police corruption, the draft resistance, drug information, and adverts for sexual contacts and services," and he garnered minor cult fame because of his raw, screaming, profane, and provocative poetry and prose in the mimeograph press.[31]

Bukowski, known as the "Skid Row Poet" because of his depiction of seedy urban landscapes, populated by the unemployed, by drunks, and by prostitutes, produced much of his poetry along 34°0′0′ (DMS) latitude as it runs through East Hollywood and the other grittier parts of Central Los Angeles (Fig. 2.3). In *My Kind of Place* (1978), Bukowski describes what it is like to be marginalized under the famous hillside sign that signifies to the world the illusory dreamland of the American experience:

---

[30] Bukowski, C., 2009. *The Roominghouse Madrigals: Early Selected Poems 1946–1966*. Harper Collins.

[31] Debritto, A., 2013. *Charles Bukowski, King of the Underground: From Obscurity to Literary Icon*. Springer.

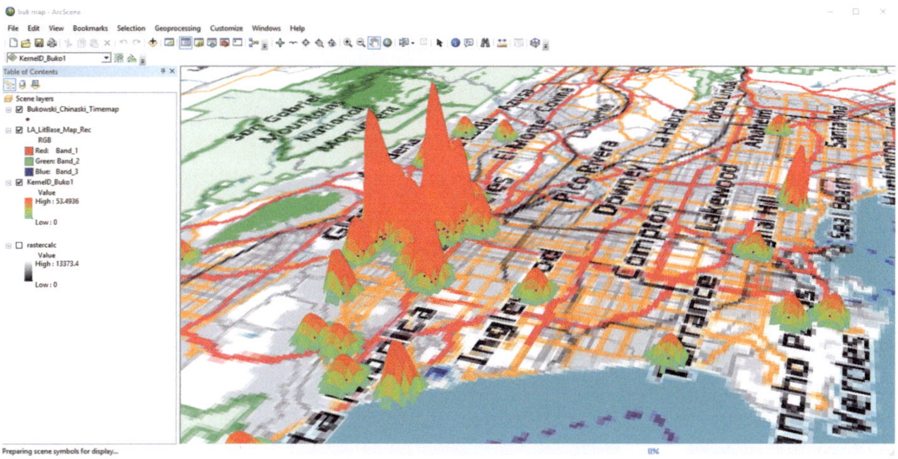

Fig. 2.4 Charles Bukowski's "Peaks and Canyons of Experience" (Charles Travis)

I can see the 'Hollywood' sign on the mountain
and I walk the streets in the late afternoons
dressed in bluejeans and a
black t-shirt.
it's warm and easy and there's not much to do.
the black whores take up most of the tables at the STAR BURGER
and I walk past ZODY'S
carrying a 6 inch switchblade in my pocket.[32]

Bukowski moved to the seaside suburb of San Pedro in 1978 (illustrated by the dip in the graph above) but lived for most of his 73 years alongside the "Skid Row Latitude" of 34°0′0′ as it bifurcated the sweltering, smog-filled belly of Los Angeles. His piece, *Nut Ward East of Hollywood*, describes navigating the milieu:

> We finished the wine and then walked down to Shakey's and drank the deep brown beer by the pitcherful and watched the old-time fights – we saw Louis get dumped by the Dutchman; the third Zale–Rocky G. fight; Braddock–Baer; Dempsey–Firpo, all of them, and then they put on some old Laurel and Hardy flicks.[33]

Bukowski stated his writing reflected literally what had happened in his life and his experience living on the bottom rung of American society. The following lines from his poem *The New Place* illustrate the "sense of place" of the lowest rung (Fig. 2.4):

the manager wears all white
has a 52 inch color tv
and sits in the garden with her

---

[32] Bukowski, C., 1978. My Kind of Place. *Wormwood Review*, 18:3. Pp. 96–97.

[33] Bukowski, C., 2008. *Tales of ordinary madness*. Random House. Pg. 31.

x-alcoholic husband and speaks
of the price of red rose potatoes.[34]

The visual topography in Fig. 2.4 illustrates the predominant locations of Bukowski's life experiences and poetic production. The highest and lowest "Peaks and Canyons" of his life was lived along the "Skid Row Latitude" of 34°0′0′ (DMS). From *Flower, Fist and Bestial Wail* (1960), his first full-length collection of poetry, to *Pulp* (1994) published after his death, Bukowski employed the lens of "Dirty Realism" to depict "the humorous, lyric, impoverished lives of prostitutes, drinkers, bums, writers, and miscreants of every description."[35] Dirty Realism explores the "belly side of contemporary life" focusing on "local details, the nuances, the little disturbances in language, and gesture" which draws our attention to the

> … unadorned, unfurnished, low-rent tragedies about people who watch day-time television, read cheap romances … drink a lot and are often in trouble: for stealing a car, breaking a window, pickpocketing a wallet.[36]

Bukowski claims that when he started reading literature, almost nothing he found "related to me or the streets or to the people around me" and as a result disavowed the "senses of place" espoused by the American literary schools of his generation: "Those Black Mountain School snobs, let them smell their own turds! The Kenyon boys, let them write their celluloid senseless inoffensive poems." Bukowski concluded "to me, the entire poetic scene seems dominated by obvious and soulless and ridiculous and lonely jackasses … from the university group at the one end to the beat mob at the other … they go from creators to being entertainers."[37] Described as the "only major post-War American writer who has denied the efficacy of the American dream," Bukowski's work is viewed as "typically individualist, anti-formal, anarchistic" in its critique of the "Protestant work ethic, American market capitalism, and how these things affect the individual and society."[38] Despite Bukowski's conviction that academics are "parasites on the cerebral who rung texts out to dry to satisfy moribund preconceptions,"[39] this sketch's deep mapping was guided by contrapuntal *distant* and *close* readings of his texts and the places of their production (typically where he happened to be living) to plot the locations of his literary experience and perceptions of Los Angeles. Engaging literary critic Walter Benjamin's cartographical aspirations and spatial sensibilities to explore the locations of Bukowski's work and life sets the stage for deeper mapping of Los Angeles in order to consider how the milieus and paths of his and LA writers like Raymond Chandler,

---

[34] Bukowski, C., 1976. The New Place. *Wormwood Review*, 16:1. Pp. 36–39.

[35] Madigan, *What Fame Is*. Ibid.

[36] Buford, B. 1983. Editorial, in Buford, B. (Ed.). *Granta 8: Dirty Realism: New Writing From America*. Granta, Cambridge. Pg. 4.

[37] Debritto, A., 2012. Writing into a Void: Charles Bukowski and the Little Magazines. *European journal of American studies*, 7:7–1.

[38] Madigan, *What Fame Is*. Ibid.

[39] Malone, A. 2003. *The Hunchback of East Hollywood: A Biography of Charles Bukowski*. Headpress/Critical Vision, Manchester. Pg. 43.

Walter Mosley, T.C. Boyle, and other writers' actual and fictive *lifeworld* paths intersect and interpenetrate to tell the stories of the city.

Dublin, Ireland: 1916 Easter Rising/2016 Commemoration

On Easter Monday, 24 April 1916, the occupation of the General Post Office (GPO) in Dublin, Ireland, signaled the intentions of a small group of republican and socialist militants to deliver a final proclamation against British rule in Ireland. The occupation led to an urban military conflict between the Irish rebels and British Army forces which left the city center of Dublin in ruins. The "Rising" was put down by April 30th and, though unpopular among the majority of Dubliners, only achieved totemic status in the Irish political imagination after the rebel leaders were summarily executed by the British. This led to a cascade of political events that inaugurated the Anglo-Irish War of 1919–1921 and the formation of the southern 29 county Irish Free State in 1922 (the Republic of Ireland was declared in 1949). The military parade past the General Post Office on O'Connell Street Dublin to commemorate the 1916 Easter Rising was reinstated in April 2006 after a hiatus of over three decades. The parade was halted in 1972 after being viewed as an incendiary act in light of the sustained political and sectarian violence of the 'Troubles' in Northern Ireland. On the ninetieth anniversary of the 1916 Rising in 2006, it was felt that a grace period of eight years from the 1998 Northern Ireland Peace Process Agreement had passed. As the national mood in the Republic of Ireland was cautious and reconciliatory, the parade was resumed (Fig. 2.5).[40]

As illustrated by the following lines from William Butler Yeats' poem *Easter 1916*, Rising commemorations during the twentieth century typically valorized the male protagonists who were executed and transformed in martyrs for Irish nationhood.

To know they dreamed and are dead;
And what if excess of love
Bewildered them till they died?
I write it out in a verse-
MacDonagh and MacBride
And Connolly and Pearse
Now and in time to be,
Wherever green is worn,
Are changed, changed utterly:
A terrible beauty is born.[41]

[40] Smyth, H. 2015. *Commemoration from Below: Mapping the Civilian Fatalities of the 1916 Rising*. Master of Philosophy Public History and Cultural Heritage Thesis (Unpublished) Trinity College, The University of Dublin.

[41] Poetry Foundation, accessed August 2017 <https://www.poetryfoundation.org/poems/43289/easter-1916>.

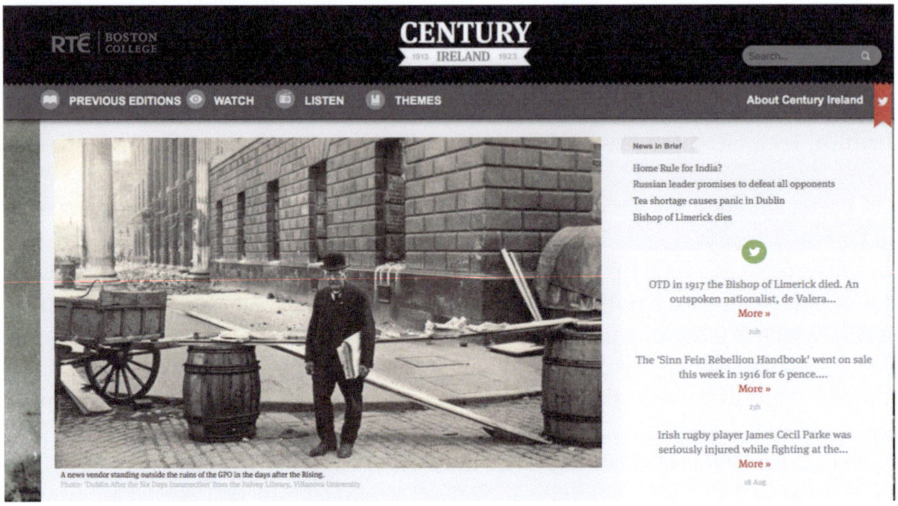

Fig. 2.5 Image of GPO. After Easter 1916 Rising from *Century Ireland* Twitter Narrative

Such sentiments contributed to official state narratives that elided the participation of other historic voices and actors. Additionally, and perhaps most troubling, such narratives contributed to the myth of blood sacrifice as a means to achieve a free Ireland and the rationale for political violence in Northern Ireland. At the behest of the Irish Dáil (parliament) commemoration committee, a minute silence was held explicitly in remembrance of all those who died during Easter Week 1916 whether they were a rebel, in the British Army, a Dublin Police, or a civilian. Accordingly, for the first time, civilian deaths caused by the Rising were publicly and politically acknowledged. However, civilians remain a footnote in the history of the revolutionary period and of the 1916 commemorative agenda (Fig. 2.6).[42]

Deep mapping the 1916 Rising/2016 Commemoration can help us understand how the absences of innocent civilians still linger in Irish historiography and provide an instructive example of how contentious public history commemorations and remembrances are mapped. The visualization in Fig. 2.6 represents the locations of where the lives of innocent civilians caught up in the urban conflict of the 1916 Rising in Dublin, Ireland, tragically ended. One of the main representational issues is to recognize that these locations are just a starting point from which to work backward and plot out the individual *lifepaths* of the deceased and gather the impressions of their *lifeworlds* (Fig. 2.7).

We could imagine the collective *lifepaths* reaching into the depths of history and in places intersecting under the three-dimensional schema of the city displayed by the 1903 *Thom's Map of Dublin* (Fig. 2.7) like the roots or branches of a tree or like the dendritic tendrils of the human brain which carry our memories, hopes, and

---

[42] Travis, C and Smyth, H. 2016. Tell the Story of Irish Public History, *Learn ArcGIS*. Esri Press. <https://learn.arcgis.com/en/projects/tell-the-story-of-irish-public-history/>.

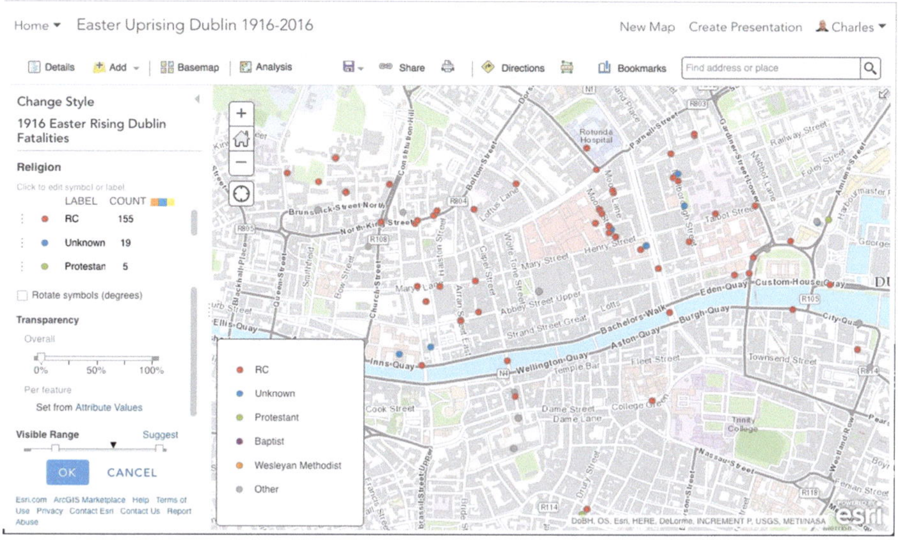

Fig. 2.6 Deaths from 1916 Rising by religious denomination (Charles Travis, courtesy of Hanna Smyth, the Glasnevin Trust and Esri)

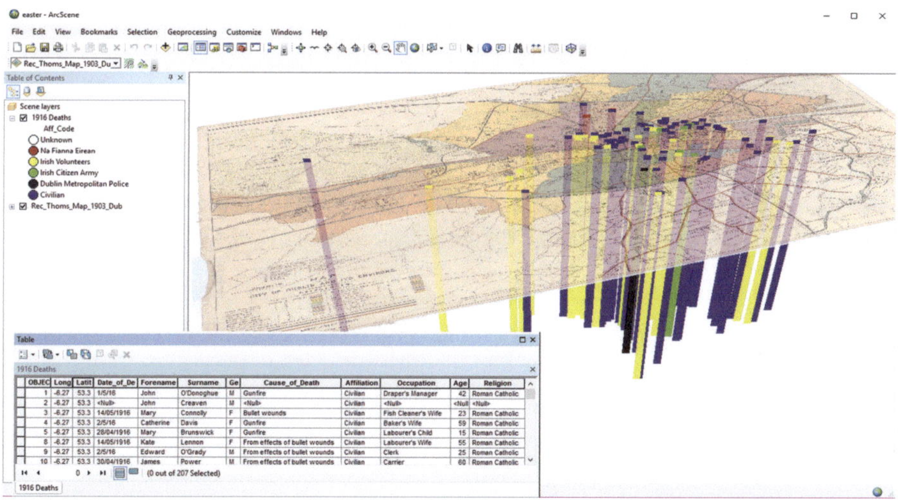

Fig. 2.7 Easter Uprising 1916 deaths (civilians in blue), deep map tendrils (Charles Travis)

fears.[43] For instance, we could link the locations of these civilian deaths—contextualized by the 1916 Rising locations and a 2016 *Twitter* narrative of the occupation of the General Post Office by the rebels and subsequent events generated by *Century*

[43] 1903 Thom's Map of Dublin image courtesy of Glucksman memorial map library, Trinity College Dublin.

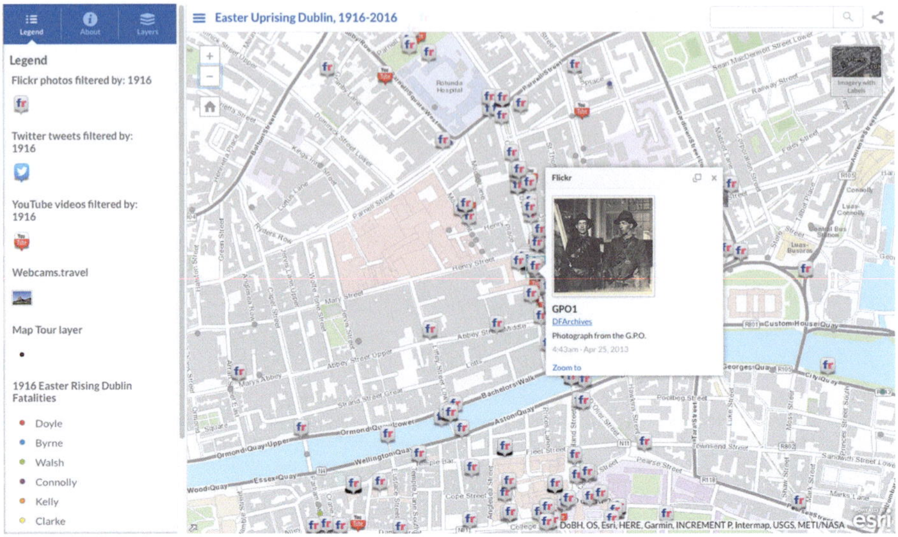

Fig. 2.8 Historical commemoration social media post from the Dublin GPO (Charles Travis)

Ireland (Fig. 2.2) to social media posts (which cluster around the 2016 Dublin Commemoration Parade route) by individuals and groups observing the centenary (Figs. 2.5, 2.6, and 2.8).

Advances in cloud computing now allow the fruits of the Internet of Things, such as social media-generated "Big Data," to be processed remotely on desktop and tablet computers, and for HumGIS practitioners, such data streams can now supply the "deep content" required to analyze patterns and trends about the "many."[44] The Rising collapsed civic life in Dublin and began the cascade which removed the city from the British Cartesian Pale, but the fate of the innocent lives caught between the forces of empire and rebellion has only recently been explored due to the political turbulence in twentieth-century Irish history and the partition of the island. But with political violence receding and the presence of a Northern Peace Process, the mapping of human experience, agency, and identity, rather than the hagiographic mapping of battles, monuments, and iconography, is now emerging as one of the themes of Irish public history (Fig. 2.9).

The 1916 Rising/2016 Commemoration deep mapping exercise illustrates that emerging human-centric GIS "knowledge systems"[45] both embrace and transcend

[44] Sui, D. and Goodchild, M.F. 2011. The convergence of GIS and social media: challenges for GIScience. *International Journal of Geographical Information Science*. 25:11, 1737–1748.

[45] Bodenhamer, D., Corrigan, J. and Harris, T.M., 2010. *The Spatial Humanities: GIS and the Future of Humanities Scholarship.* Bloomington: Indiana University Press; Dear et al. 2011. *Geohumanities: Arts History, Text at the Edge of Place* (Routledge); Travis, C. 2015 *Abstract Machine: Humanities GIS* (Esri).

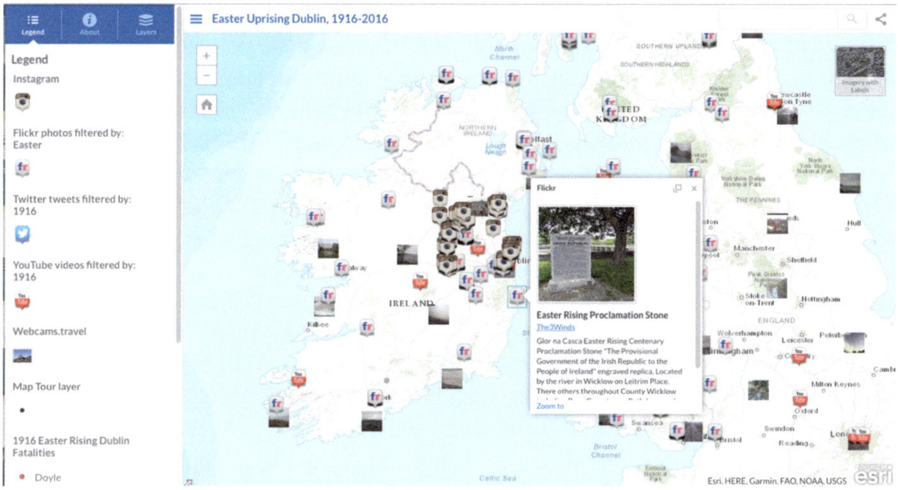

Fig. 2.9 2016 Commemoration social media post

Cartesian-based geospatial technology perspectives which frame the relation between "place" and historical events as a skeletal "geometry with names."[46] By harnessing nontraditional geo-data sources and employing a palette of affect and subjective perception, Spatial Humanities GIS can paint the colors and plot the vagaries of human experience and behavior, introducing a STEAM bridge between urban planners and engineers, public historians, and literary, digital, cultural, and media scholars.

Conclusion

> To educate the image-making medium within us, raising it to a stereoscopic and dimensional seeing into the depths of historical shadows. The words are from Rudolf Borchardt's *Epilegomena zu Dante*, v. 1. [Berlin 1923] pp. 56–7. (Walter Benjamin, *The Arcades Project* (N 1, 8), 458)

In *De Antiquissima Italorum Sapientia* (1710), the Italian eighteenth-century political philosopher and historian Giambattista Vico critiqued his contemporary René Descartes' methods for studying "civic life," stating that trying "to import the geometric method into practical life" was like "going mad by means of reason" and one "would march straight ahead through the infractuous course of life as though desire, temerity, occasion, fortune did not rule in human affairs."[47] In contrast to

[46] Olsson. G. 2007. *Abysmal: A Critique of Cartographic Reason*. Chicago: The University of Chicago Press. Pp. 137–138.

[47] Vico, G. 2010 [1710]. *De Antiquissima Italorum Sapientia* (On the Most Ancient Wisdom of the Italians) Trans. J. Taylor. New Haven and London: Yale University Press. p. 11.

traditional Cartesian perspectives and approaches to studying cities and civic life, declaimed by Vico, the Spatial Humanities GIS urban case sketches in this chapter show how "three key referencing systems -space, time and language-might be engineered in such a way that changes in one ripple into the others."[48] GIScientist Michael Goodchild notes that over the past two decades, GIS has gradually transformed into a type of media in which "place" perspectives have gained ascendancy.[49] The confluence of the Internet of Things, humanities Big Data archives, and contemporary social media activity presents scholars who adopt HumGIS approaches with incomprehensibly large seas of information to navigate, explore, harvest, and survey. The social theorist Bruno Latour informs us: if you "change the instruments ... you will change the entire social theory that goes with them."[50] The traditional Cartesian perspective of many contemporary geospatial technology platforms adopts an "objective" separation between the perceiver (subject) and the perceived (object). In contrast (to borrow from the Czech author Milan Kundera) in phenomenological deep mapping perspectives: "man and the world are bound together like the snail to its shell: the world is part of man, it is his dimension, and as the world changes, existence (*in-der-Welt-sein*) changes as well."[51] Humanistic geographer Anne Buttimer explains: "world to a phenomenologist is the context within which consciousness is revealed. It is not -a mere world of facts and affairs, but ... a world of values, a world of goods, a practical world. It is anchored in a past and directed towards a future; it is a shared horizon, though each individual may construe it in a uniquely personal way."[52] In many ways, language and perceptions "bracketed" by literature, historical, and cultural documents in addition to discursive and visual forms of social media reveal multiple slices of idiosyncratic *lifeworlds* constellating across time and space, forming and reforming into unique and contingent "senses of place." Buttimer suggests: "if people were to grow more attuned to the dynamics and poetics of space and time, and the meaning of milieu in life experience, one could literally speak of the [...] personality of place which would emerge from shared human experience and the time-space rhythms deliberately chosen to facilitate such experiences."[53] Phenomenological and cinematographic perspectives applied to Spatial Humanities GIS deep mapping intimate Gunnar Olsson's observation on GIS and remote sensing:

[48] Corrigan, J. 2010. Qualitative GIS and Emergent Semantics. *In*: D. J. Bodenhamer, J. Corrigan, T.M. Harris, eds. *The Spatial Humanities: GIS and the Future of Humanities Scholarship*, Bloomington & Indianapolis: Indiana University Press, 76–88. Pg. 85.

[49] Goodchild, M. F. 2011. Formalizing place in geographic information systems. In L. M. Burton, S. 540 P. Kemp, M.-C. Leung, S. A. Matthews, & D. T. Takeuchi (Eds.), *Communities, neighborhoods,* 541 *and health: Expanding the boundaries of place.* New York: Springer. Pp. 21–35.

[50] B. Latour. 2009. Tarde's idea of quantification. *The Social after Gabriel Tarde: Debates and Assessments*, ed. M. Candea, Routledge, London, 145–162. Pg. 9.

[51] Kundera, M. 1988. *The Art of the Novel.* London: Faber and Faber. Pg. 35.

[52] Buttimer. *Grasping the Dynamism of Lifeworld.* Pg. 246.

[53] Ibid., 290.

> […] for what is that type of mapping at a distance if not a human activity located in the interface between poetry and painting? What is a satellite picture if not a peephole show, a constellation of signs waiting to be transformed from meaningless indices into meaningful symbols?[54]

Spatial Humanities GIS approaches proscribe certain types of hermeneutic, cultural, and affective deep mapping representations and analyses, in contrast to traditional GIScience "paint by numbers" dataset framings and methodologies. In conclusion, as Buttimer notes in this chapter's epigraph, "behavior in space and time [is like] the surface movements of icebergs, whose depths we can sense only vaguely."[55] Spatial Humanities GIS deep mappings of the city holds the potential to illuminate and represent past and present "senses of place" rooted in our human condition's "latent substratum of experience."[56] In the future, an "engaged pluralism" born from the cross-pollination of STEAM-framed urban, planning, engineering, literary, historical, cultural, geographical, and digital humanities research methods "can become the norm," offering a counterpoint to the "abstract vision for the future" framed contemporarily by positivistic and quantitative approaches.[57]

[54] Olsson, *Abysmal: A Critique of Cartographic Reason*. Ibid., 137–138.

[55] Buttimer, *Grasping the Dynamism of Lifeworld*. 287.

[56] Ibid.

[57] Kwan, M.P. and Schwanen, T., 2016. *Geographies of mobility*.

Chapter 3
STEAM Approaches to Climate Change, Extreme Weather and Social-Political Conflict

Francis Ludlow and Charles Travis

Abstract Climate history and historical climatology are closely evolving fields that aim (1) to reconstruct past climatic conditions and (2) examine the societal impacts of climatic changes. Such research can be characterized as a multi-disciplinary STEAM (Science, Technology, Engineering, Arts, Humanities and Mathematics) frontier that thrives under multidisciplinary collaboration. This chapter presents two case studies pertaining to each aim, linking palaeoenvironmental data from ice-cores and tree-rings with climatic and societal information preserved by the long Irish tradition of annalistic record keeping between the fifth or sixth to seventeenth centuries. The product of this recording survives today in texts known collectively as the "Irish Annals", which provide detailed time series of extreme weather experienced in Ireland. The first case study revisits work that employs the Irish medieval record of severe cold weather together with Greenland ice core sulphate records to reveal a persistent winter-season climatic impact from explosive volcanism at Ireland's climatically sensitive northeast Atlantic location. This result complements evidence of spring-summer (i.e., growing season) volcanic climatic impacts identified from tree-rings, and furthers our understanding of the potential impacts of the next big eruption or of geoengineering implemented via the stratospheric injection

Francis Ludlow is a Marie Skłodowska-Curie Individual Fellow, 2016–2018, at the Department of History, School of Histories and Humanities, Trinity College Dublin. This publication represents a contribution to the "Historical Dynamics of Violence, Conflict and Extreme Weather in Medieval Ireland" (CLIMCONFLICT) project. This project has received funding from the European Union's Horizon 2020 research and innovation programme under the Marie Skłodowska-Curie grant agreement No 709185. Francis acknowledges additional support from the Trinity Centre for Environmental Humanities, Trinity College Dublin, and input from the Volcanic Impacts of Climate and Society (VICS) Working Group of PAGES.

F. Ludlow (✉)
Trinity College, The University of Dublin, Dublin, Ireland
e-mail: ludlowf@tcd.ie

C. Travis
Department of History, University of Texas at Arlington, Arlington, TX, USA

Trinity Centre for Environmental Humanities, Trinity College, The University of Dublin, Dublin, Ireland
e-mail: charles.travis@uta.edu

© Springer International Publishing AG, part of Springer Nature 2019

A. de la Garza, C. Travis (eds.), *The STEAM Revolution*,
https://doi.org/10.1007/978-3-319-89818-6_3

of sulphur dioxide. The second case study compares Irish annalistic evidence of violence to evidence of drought gleaned from Irish oak tree-ring growth widths. This exercise reveals suggestive linkages between extreme weather and violence, operating most obviously (but likely not solely) via the pressures of scarcity induced resource competition. This comparison also shows that medieval Irish society was not a passive victim of extreme weather, with a range of coping strategies available to restore order. Such knowledge is critical at a time when an increased risk of conflict arising from anthropogenic climatic change is regarded by many scholars and policymakers as a key security issue. Suggestions that human-induced climate change may act to catalyze contemporary or future violence and conflict remain controversial, however, with data shortages and the complexity of societal "pathways" that may connect climate to conflict presenting major research barriers. STEAM approaches that combine different methodologies and evidentiary bases to facilitate the examination of multiple historical climate parameters and a broad range of conflict typologies are thus essential to identifying the range of possible correlations between climate, violence, and conflict, and to resolving the complex pathways underlying observed correlations. Both case studies presented here illustrate how multidisciplinary STEAM approaches, well represented by climate history and historical climatology, can meaningfully inform contemporary debates by revealing how humanity has been influenced by past environments.

Keywords STEAM · Climate history · Historical climatology · Extreme weather · Explosive volcanism · Climate-conflict linkages · Irish annals · Medieval chronicles · Medieval history · Ireland · Mitigation

Introduction

An increased risk of conflict arising from human-induced climatic changes that increase the frequency and severity of extreme weather events and exacerbate natural resource inequalities is regarded by many scholars and policymakers as a critical security issue (Barnett and Adger 2007; Steinbruner et al. 2012; Schleussner et al. 2016). This is, perhaps, nowhere more pressing a concern than the contemporary Middle East, but suggestions of links between climate and conflict often prove controversial. This is exemplified by the unresolved debate about the role that a multi-year (2007–2010) drought may have played in triggering the 2011 Syrian uprising (e.g., De Châtel 2014; Kelley et al. 2015). The origins of a situation such as that seen in Syria will be inevitably complex, and context is king. In this case, living under dictatorial governance with markedly unequal access to resources is, for example, unlikely to be a good recipe for long-term social stability. But this does not preclude a contribution from climate. Indeed, these same socioeconomic conditions can plausibly predispose a society to violence following extreme weather by limiting the available coping strategies and diminishing adaptive capacities. Yet despite the importance of such questions, the degree to which climatic conditions may influence conflict remains under-researched (Barnett 2003). Studies continue to yield contradictory or contested results (e.g., Fjelde and von Uexkull 2012; Hsiang and Burke 2014; Buhaug et al. 2014; Hsiang et al. 2014), with data limitations and the complexity of societal "pathways" that may connect climate to conflict presenting major ongoing barriers to understanding (Scheffran et al. 2012). Because the role of climate as a trigger or contributor to conflict will in all cases be mediated by the prevailing social and political background, historians working from STEAM approaches can contribute by supplying nuanced and longer-term perspectives.

To identify and characterize the full range of possible linkages between climate and conflict, multiple climate parameters must be examined against a broad range of conflict typologies (e.g,. large- vs. small-scale, organized vs. spontaneous). It is also crucial to understand that statistical correlations do not by themselves reveal causation. A critical further challenge is, therefore, to resolve the complex climate-conflict pathways underlying any observed correlations and to identify the socio-economic contexts that diminish adaptive capacities and increase vulnerability to climatic changes, thereby heightening the risk of violence and conflict (Adano et al. 2012). Multidisciplinary perspectives and approaches of the type embodied by STEAM are crucial for achieving such progress and can be very broadly applied to the long lens of history to examine how humanity has influenced, and been influenced by, the environment (e.g., Hassan 2000; Campbell 2010; Izdebski et al. 2016). Historical climatology and climate history are closely related fields, both of which aim with varying degrees of emphasis to understand or reconstruct past climatic conditions from the evidence of written sources and to examine how past climatic conditions influenced society (with the latter aim being the more dominant emphasis of scholars who identify as climate historians rather than historical climatologists, though this distinction is not formal and far from absolute (see, e.g., Pfister (2010)). Both fields can be clearly seen as operating within an interdisciplinary STEAM frontier. This chapter will discuss the potential and challenges of research in historical climatology and climate history, drawing upon case studies illustrative of the two main interrelated strands of work in this frontier field. Adopting a didactic framing, the first strand can be usefully seen as a case of history (its sources and methods) informing science, taking the example of written evidence of the climatic impact of volcanism preserved within medieval Irish chronicles. The second can be seen as a case of (natural) science informing history, taking the example of tree-ring-based evidence of past weather extremes informing the study of historical societal vulnerability and response, including violence and conflict, to sudden environmental hazards.

History Informing Science

Climate historians and historical climatologists have been increasingly active in utilizing the documentary heritage of Europe and Asia, and more recently the Middle East, Africa and the Americas. Relevant sources used to reconstruct climatic conditions prior to the twentieth century when instrumental meteorological records become increasingly scarce include weather diaries, personal diaries, ships' logbooks, early print journals and newspapers (Pfister et al. 2002; Brázdil et al. 2005). Historical climatology has a long pedigree, with recognizable works in this field published as early as the 1920s (e.g., Brooks 1928) and antecedents among compilers of historical catalogues of weather back to the eighteenth century and beyond, one example being John Rutty's 1770 catalogue of weather conditions, which he related statistically to instances of disease in Dublin city (see also Britton (1937)).[1] Since the 1990s in particular, the field has achieved an increasing prominence, and multiple documentary-based reconstructions now stretch back to the sixteenth century, primarily for Europe and Asia. Even with the rich documentation from these regions, however, the eleventh century represents a major barrier (as noted by Lamb (1995)) because of a relative paucity of earlier sources and an often unfair perception that many of those that do exist are not of credible use, being imprecisely dated, exaggerated or deemed otherwise unreliable as non-contemporary copies of now-lost originals. Even between the eleventh and sixteenth centuries, many historical climate records remain untapped, when they are more fragmentary and more alien in form, language and content to modern readers. This absence of documentary-based reconstructions has hampered exploration of the existence and character of proposed large-scale climatic phases such as the Medieval Warm Period (now more often termed the Medieval Climatic Anomaly) c.900–1300, as well as an understanding of the nature of the transition from the Medieval Warm Period into the subsequent Little Ice Age (c.1350–1850).

Early works of historical climatology were criticized on the grounds of insufficient assessment of the historical reliability of source material (e.g., Bell and Ogilvie 1978; Ogilvie and Farmer 1997). While the need to assess the reliability of all documentary evidence employed in climatic reconstruction is now noted almost ubiquitously in the historical climatology literature, most published reconstructions can afford only limited room to detail examples of the assessment process. What is often primarily stressed can be described as the "contemporaneity criterion" (Ludlow 2010), in which only sources deemed strictly contemporary to the events they record are to be considered for use in climate reconstruction. While a focus on such sources is appropriate and logical, it can also serve to cut out a lot of potentially valid material, especially for earlier periods when we depend frequently upon later copies of once-contemporary sources. Yet these materials *can* be assessed for reliability and used in studies of medieval climate (e.g., McCormick et al. 2007; Gao et al. 2016). The lack of detailed historical methodology in most published climate reconstructions does not, however, help guide researchers interested in using the pre-1100 material. A mainstay of the available material pre-1500 and especially pre-1100 are annals and chronicles, which generally comprise annual lists of events deemed important by the original authors or later copyists. These sources have not generally been regarded as being of the same utility as others such as weather diaries that

[1] John Rutty, 1770. *A chronological history of the weather and seasons, and of the prevailing diseases in Dublin: with the various periods, successions and revolutions, during the space of 40 years: with a comparative view of the difference of the Irish climate and diseases and those of England and other countries*. London: Roberts and Roberts.

record weather conditions on a continuous daily or weekly basis (e.g., Le Roy Ladurie 1971), but a growing appreciation exists for the record of extreme weather captured by annals and chronicles (e.g., Guillet et al. 2017; Ludlow 2017). This is because few natural proxies such as tree rings can detail the occurrence of singular extreme events, and the historical record thus allows an examination of the relationship between their occurrence and longer-term variations in average climate (e.g., as can often be more clearly registered by natural proxy archives). As extreme events are inherently rare, the long temporal span of annals and chronicles, often covering multiple centuries, also provides a great advantage in attempting to characterize the behaviour of extreme weather under a changing climate, something that is of major importance in predicting likely changes in the frequency, seasonality and severity of extreme events as global warming proceeds.

A little-known fact is that Ireland boasts the earliest Christian annalistic record in Europe (Dunphy 2010). Although some debate persists concerning the start date of contemporary annalistic record keeping (or chronicling) in Ireland, as well as how representative the content in the surviving manuscripts is of this early Irish chronicling, it seems likely that the extant manuscripts contain genuine material from at least the sixth century, and potentially even the fifth, though certainly interspersed with materials interpolated during later copying and materials borrowed from continental chronicles (McCarthy 2008). Early chronicling of historical events in Ireland began at first in major Irish monastic foundations, such as on the Scottish island of Iona, but by the thirteenth century, the professional hereditary historians in the employ of the Gaelic nobility began to dominate the role of chronicler. The Irish tradition of chronicling ceased only in the seventeenth century when the manuscripts and their authors were targeted by the New English colonial government intent on stamping out Gaelic culture and history (Smyth 2006; McCarthy 2008). But in providing over a millennium of yearly chronicling that included a focus on extreme weather, the Irish Annals (as medieval Irish annalistic texts are collectively known) provide a chance to examine the incidence of the full spectrum of extreme weather events (temperature, precipitation and windiness) at a climatically sensitive location of Northwest Europe back into the first millennium, a period for which Ireland avoided most of the negative impacts of the fall of the Western Roman Empire that resulted in a comparative disruption of societies, and hence a relative paucity of documentation, elsewhere in Europe at this time.

Despite this, an influential and broadly valuable review of temperature reconstructions for the past 2000 years under the aegis of the US National Research Council (2006, p. 39) reflects a prevailing (if gradually diminishing) perception when asserting that "there are…weather records preserved in Irish and Norse annals back to the middle of the first millennium A.D., but their dating is imprecise and descriptions of weather and climate often are exaggerated". No supporting evidence or citations are provided here, and the important work of McCarthy (1998, 2001, 2005, 2008) in correcting chronological errors in the early Irish annalistic chronology has been overlooked. It is certainly true that all documents require careful assessment and some are challenging. Written sources may exaggerate or fabricate information and misunderstand or misreport phenomena witnessed. Extreme events are also, as stated above, rare by definition, and thus the uncritical inclusion of even small numbers of events based on unreliable information can notably bias climate analyses. Yet historians excel at assessing the reliability of the evidence in written sources, doing so by inhabiting the mind-set of authors from far removed times and places and by understanding nuance and the importance of the context in which any document is written (MacNeil 2000). This provides a means to rigorously assess and employ the evidence of even challenging written sources in climatic reconstruction, exemplifying just one way that history (and historians and textual scholars) can inform work in the natural sciences. The following section thus expands upon the work of Ludlow et al. (2013) to first demonstrate the importance of assessing the credibility of the severe cold weather reported in medieval Irish annals, before providing an example of the utilization of this record of cold to better understand the climate history of the Northeast Atlantic region of Ireland.[2]

Historical Climatology and the Irish Annalistic Record of Cold

A vellum manuscript page from the famous *Annals of Ulster* is featured in Fig. 3.1, in which the manuscript's text covers the years 852–858, with each year presented in distinct paragraphs. Work on the compilation of this manuscript began in 1498 CE, representing one of the most comprehensive attempts of late medieval Irish scholars to compile an annalistic history of Ireland by drawing upon the multiple texts that then existed, many of which are now unfortunately lost (McCarthy 2008). Written at this point in a macaronic mixture of Latin and Irish, the first line of each paragraph denotes the start of each year's record using distinctive chronological notation, chiefly an enlarged "Kl" that signifies the Kalends of January (i.e., January 1st), with a corresponding Anno Domini date in Latin numerals.[3] As with other medieval annalistic records, the text preserves a list of yearly events deemed notable by the scribes that first compiled it, for these years almost certainly working in major Irish monastic centres, but also those later copyists and compilers that iteratively preserved these earlier records in later manuscripts through the centuries. Despite the loss of many manuscripts and their texts, Ireland is still abundantly endowed with these sources, which survive in approximately 1.14 million words in 20+ major and minor texts, and provide considerable coverage of extreme weather conditions and related societal impacts, for all seasons (Ludlow 2010; Hickey 2011; Mitchell 2011; Ludlow et al. 2013).

---

[2] Portions of text are adapted here from Ludlow (2012), with permission of the publisher, Geography Publications.

[3] The text of this manuscript variously employs additional chronological notation for different years. The seminal work of McCarthy (2005, 2008) and citations provided therein provide a full treatment of this aspect of the *Annals of Ulster* and other extant medieval chronicles from Ireland.

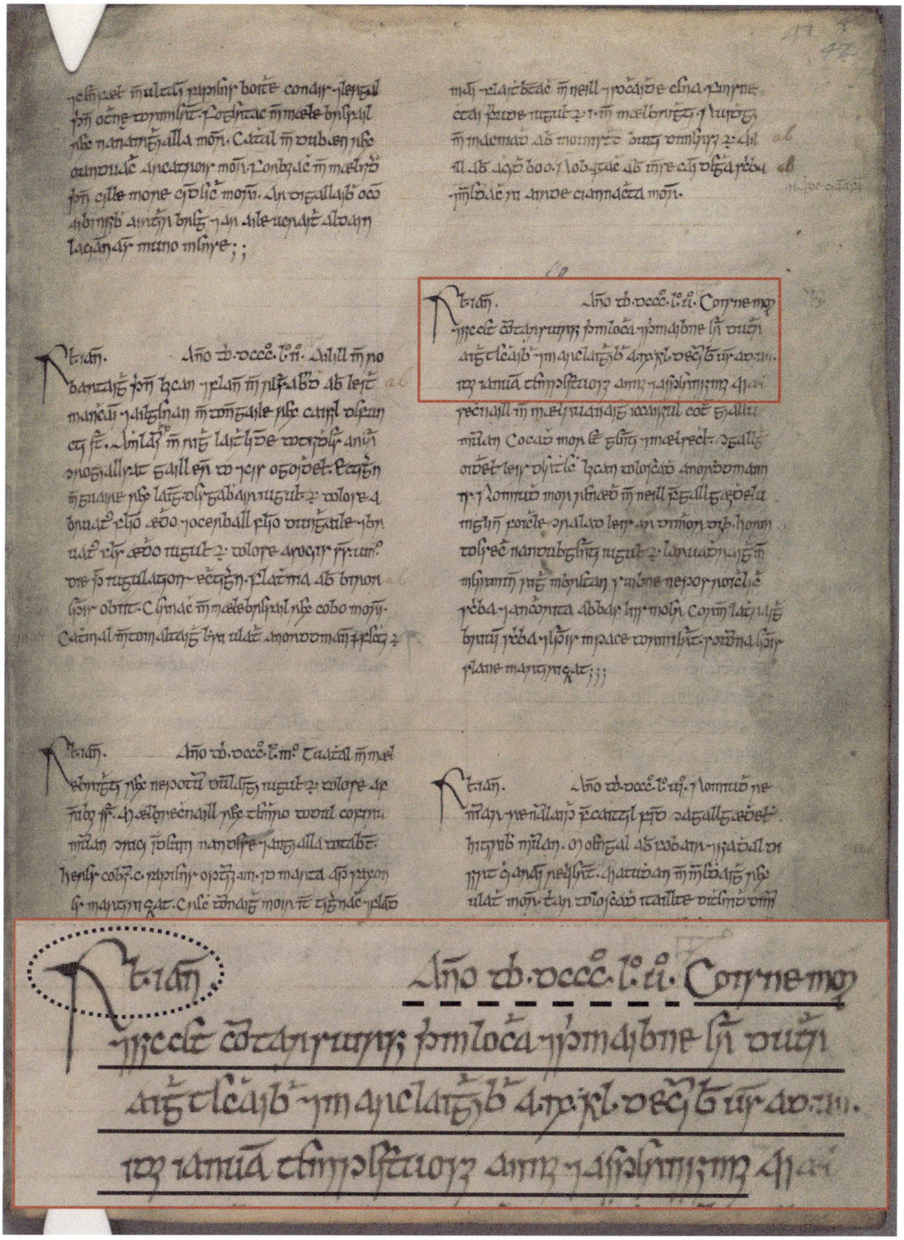

Fig. 3.1 Vellum page (H 42r b) from TCD MS1282, the Annals of Ulster, covering the years 852 to 858 CE. In the inset, the encircled notation reads "K1. Ianair", representing early medieval Irish chronological notation denoting the "Kalends of January", i.e., the first day of January, and marking the start of this year's list of entries. The dashed underlined text represents the Anno Domini date in Latin numerals, while the solid underlined text represents the first historical entry for this year (856), reporting the severe cold weather cited in the main text. Image reproduced with permission from the Board of Trinity College Dublin.

The inset in Fig. 3.1 highlights a characteristic example of a "cold event" reported in the *Annals of Ulster*, being the first report for the year 856.[4] This states in translation: "There was much ice and frost so that the principal lakes and rivers of Ireland could be crossed by people on foot an on horseback from the ninth of the Kalends of December [i.e., 23 November 855] to the seventh of the Ides of January [i.e., 7 January 856]. A tempestuous and harsh year."[5] Conditions such as these are clearly severe for Ireland's mild maritime climate (Sweeney 1997; Graham 2004). Christian Pfister and colleagues (Pfister et al. 2002, pp. 6–7) have argued that the medieval authors who documented extreme weather were "well aware that their narratives included a subjective bias" and hence "referred to physical and biological proxy indicators that were known to be more objective yardsticks and that could be compared over time." The Irish annalists made similar efforts, as per the reference to the loading-bearing strength of the ice in 855/856. A later example from the *Annals of Ulster* for 1115 further illustrates this, whilst also testifying to the deleterious impacts that extremes of cold could have upon medieval Irish society and the biosphere:

> Extremely bad weather in the form of frost and snow from the fifth of the Kalends of January [1 January 1115, Julian Calendar] to the fifteenth of the Kalends of March [15 February], or a little longer, and it inflicted slaughter on birds and beasts and men, and from this great want arose throughout all Ireland, and particularly in Laigin [Leinster].

Here the annalist conveys the severity of the extreme weather by detailing its impacts, their geographical scale, and, most objectively, by quantifying the duration of the cold. Such detail is highly suggestive of contemporary observation, even if this report now survives only in a text compiled several centuries later. Also notable in this report is the annalist's explicit acknowledgement of uncertainty (and implicitly of subjectivity) over duration, with cold conditions potentially continuing for an undefined period, if only "a little longer" after the stated end date. Despite their efforts, therefore, even contemporary annalists could not be completely objective reporters of weather conditions, nor could they be fully aware of all relevant events within their spheres of geographical or thematic interest. If criteria were ever formally established to judge the relevance for reporting of extreme events, they were inevitably less defined and more intuitive than historical climatologists would prefer (Ludlow 2012), and almost certainly included considerations of relevance other than an event's meteorological severity. Such considerations might include an event's religious significance or the political and propagandistic value of its interpretation as a reflection of divine judgement or punishment (Ludlow 2010). Copyists and later

[4]The Anno Domini date provided by the manuscript identifies this year as 855, but is identifiable as the year 856 when applying the chronological corrections of McCarthy (2005). Despite being noted in 856, the event itself is likely however to have begun in the 855 (i.e., the 855/856 winter season) based upon the description of the cold event's duration in the text.

[5]Translated quotations from the various Irish annalistic texts are, unless otherwise stated, drawn from the online hypertext versions hosted by the Corpus of Electronic Texts (CELT) website, at www.ucc.ie/celt (accessed 30 November 2017). See the citations associated with each hypertext for the original editors and translators. Dates cited from the Irish Annals are given in the Julian Calender, and always employ the chronological corrections of McCarthy (2005).

compilers will have similarly filtered their source materials and made decisions to include or exclude earlier reported extreme events according to many considerations of relevance. While such factors must be recognized in using the Irish material in any climate analyses, they do not in themselves constitute good reasons to *a priori* dismiss the evidence of these sources.

Inflation of Event Numbers from Uncertain and Duplicate Material

As a general principle, understanding the historical context in which documents were written facilitates the identification of potentially dubious material where motives other than the genuine recording of events can be discerned (MacNeil 2000). The basic unit of information in medieval annals are individual written entries listed under each year that record individual or sets of linked events or phenomena. If all entries referring to cold weather for Ireland, or reporting phenomena and conditions from which severe cold may be inferred, were used uncritically to derive a chronology of "cold events", 83 may be identified in total. We define cold events here as any apparently contiguous epsiode of ongoing cold weather, which may in some cases cross between calender years, as per the above-cited cold conditions during the 855/856 winter season. Upon critical scrutiny, however, 18 (or 27.7%) of these potential events can be deemed of uncertain reliability. One example is a report for 1022 found in the seventeenth century *Mageoghagan's Book* (also known as the *Annals of Clonmacnoise*), which states that "There fell a great wonderful snow at this time [unspecified] before the battle of Sleive Grott [Galtee Mountains, Munster]" (Murphy 1896). Although the snow is described as "great [and] wonderful", perhaps suggesting a notable volume, it appears equally likely that this phrasing has been used to signal that the purported snowfall is to be interpreted by readers as a portent or omen of the subsequent battle. Furthermore, there is reliable evidence for severe hail showers (often described in Irish as "cloch shneachta", literally "stone snow") in summer this year from other annalistic sources (Ludlow 2010). If there is any historic reality to this report from Mageoghagan's Book, it may well therefore involve a severe hail shower that has been repurposed or reinterpreted as snowfall at a presently unknown point in the chain of transmission between a potentially contemporary observation and the report as now found in *Mageoghagan's Book*. Reports such as this are certainly valuable in understanding how weather extremes may have been perceived historically and how they might be exploited based upon these perceptions, but they must clearly be used cautiously in climate analyses. It is also important to note that the 18 potentially unreliable events are not distributed evenly through time (Fig. 3.2a), thereby more severely distorting trends in apparent cold event incidence through their concentration in particular periods, such as that before 750 when potentially unreliable events constitute 53.8% of the total identified.

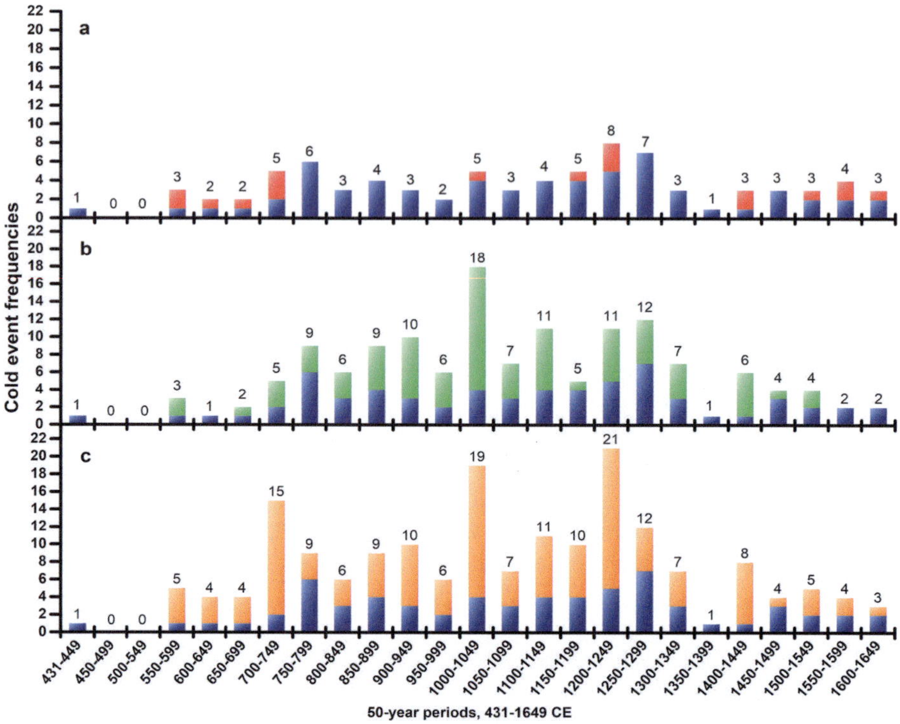

Fig. 3.2 Inflation of cold events from uncertain material and duplication. Data are arranged in 50-year periods from 431 to 1649, excepting the shorter first period. Blue bars in all three panels represent the 65 cold events deemed reliable. Stacked red bars in panel (**a**) represent potential inflation to event counts if additional events are derived from entries deemed upon critical scrutiny to be of uncertain reliability. Stacked green bars in panel (**b**) represent the potential inflation if duplicate reports of the 65 cold events are included. Orange bars in panel (**c**) represent the combined potential inflation if duplicate reports of the 65 reliable events alongside reports of uncertain events *and* their duplicates are included

As already noted, Irish annalists often copied material from older sources, thereafter extending their texts by recording contemporary events. This was necessary to preserve valuable material recorded on fragile or otherwise endangered manuscripts and has ensured that a large volume of records have survived from older sources, the originals of which are now often lost. A consequence of this copying is that many surviving sources share common ancestry and thus duplicate older entries, including those reporting weather. Details may be altered accidentally or deliberately (e.g., by abbreviation or paraphrasing) when copying sources, and chronological errors can be introduced, with discrepancies in the years under which these entries are placed in the surviving copied sources. These factors combined can effectively disguise duplicate entries, but careful assessment of similarities in orthography and phraseology allows the identification of duplicate entries (Greetham 1992; McCarthy

2005; Ludlow 2010, 2012). This must be carried out on the original Irish or Latin text, rather than using translations that can serve to further disguise duplicate material. The above work is also aided by applying chronological corrections to events reported in the Irish Annals, based upon calibration to historical and astronomical events of known date (e.g., McCarthy and Breen 1997a,b; McCarthy 2005, 2008), thus realigning duplicate entries under the correct years. Duplication can also arise where unrelated sources have independently and contemporaneously recorded events. A total of 142 entries are deemed to contain reliable information on the occurrence of cold or phenomena from which cold can be inferred. From these 142 entries, 65 reliable cold events can be identified, as noted previously. If, however, for illustrative purposes, all duplicate entries were to go unrecognized as such and a distinct cold event was thus inferred from each entry, an additional 77 events would be identified, leading to an 118.5% inflation of event totals (Fig. 3.2b). As with the additional events that might be inferred from uncertain entries (as per Fig. 3.1a), inflation from duplication is not evenly distributed and alters the apparent trend in cold event incidence through time. Entries containing unreliable material can, moreover, also be duplicated. If events inferred from these entries *and* their duplicates were also included, a further addition of 39 potential events would result, for a combined count of 181, or a maximum inflation of 178.5% beyond the 65 events in fact deemed reliable and without duplication (Fig. 3.2c). Although these figures present a hypothetical worst-case scenario, they serve to demonstrate the importance of careful assessment of source material prior to its use in climate analyses. Indeed, it is possible to go further. An assumption in the above discussion has been the possibility of fully accurately determining the reliability or otherwise of particular historical reports. In practice, this may not always be the case, and several ranking systems have thus been proposed that recognize uncertainty in the assessment of reliability, which may arise from incomplete information or inevitable subjectivity on the part of the assessor, as well as uncertainity in the identification and interpretation of phenomena described in historical texts (for cases in which a text is partially damaged or the phenomenon in question was genuinely but poorly or cryptically described by the original author). Such ranking systems (e.g., Ludlow 2010; McCormick et al. 2012; Kostick and Ludlow 2015, Sigl et al. 2015) absolve the assesor from making stark judgments of reliability or unreliability, and instead allow a more flexible and nuanced assessment.

More generally, in returning to the Irish Annals, it should be noted that despite the existence of apparently unreliable material, the Irish annalists could not as a rule have been unmotivated or inaccurate observers and reporters of natural phenomena. Whatever the criteria employed for judging the relevance for reporting of weather events, these maintained a sufficient coherence to ensure the creation and preservation of a rich and temporally extensive record. If all 313 known Irish annalistic reports of relevant meteorological and related phenomena and conditions were distributed evenly across the years 431 and 1649, they would amount to one reported phenomenon every 3.9 years (Ludlow 2010, 2012). While many influences acted to render a fraction of their reporting unreliable to one degree or another, other influences clearly existed to temper this, including the very format, strictures and conventions of the annalistic genre. The annalist who for 1492 reported in the *Annals of Ulster* on the occurrence of a hot summer cannot have been a disinterested participant in what he was reporting or copying when remarking (fully accurately on the basis of earlier annalistic content) that "a year and twenty [had elapsed] since the hot Summer before". Nor could he have been unmindful of his role in carrying forward a tradition of chronicling that spanned a millennium.

Ice-Core Records of Explosive Volcanism and the Irish Annalistic Record

Explosive volcanic eruptions have diverse societal impacts (De Boer and Sanders 2002; Oppenheimer 2011), and the 2010 eruption of Eyjafjallajökull, Iceland, which paralysed transatlantic air traffic, is a salutary reminder of the vulnerability of modern technologically advanced societies even at considerable distances from the volcanoes themselves. As well as hazards from volcanic ash or tephra, explosive eruptions can pose a societal challenge as a major cause of severe short-term climate change, as materials spewed into the atmosphere during an eruption can set in motion atmospheric changes that in turn impact climate. The basics of how volcanic eruptions impact climate are now reasonably well known (Robock 2000; Cole-Dai 2010). Large explosive eruptions loft sulphur dioxide gas (SO_2) into the stratosphere where it oxidizes to form reflective sulphate aerosol particles. These aerosols act, on average, to reduce global surface temperatures and precipitation by decreasing the amount of solar radiation reaching the Earth's surface and thereby reducing temperatures and the energy available to drive evaporation from waterbodies, as well as "dynamical" impacts on major patterns of atmospheric circulation (such the summer monsoon winds) adjusting to the Earth's altered radiation balance in cases where the loading of sulphate aersol particles is biased towards a particular hemisphere, as tends to occur after major eruptions in high-latitudes (e.g., Icelandic eruptions).

While the average global response to major explosive eruptions (particularly tropical eruptions) may now be well known, the response at regional levels may be considerably more complex because of interactions with local climatic characteristics, topography and the proximity to oceans (Robock 2000; Fischer et al. 2007; Cole-Dai 2010). The impacts of apparently similar eruptions can also be very different, depending upon the season of eruption (Kravitz and Robock 2011) and the prevailing state of major modes of climate and atmospheric variability, such as the North Atlantic Oscillation (NAO) or El Niño-Southern Oscillation (ENSO) (e.g., McGregor and Timmermann 2011; Pausata et al. 2015; Stevenson et al. 2016). Our knowledge of the impact of eruptions also comes mainly from the comparatively small number that have occurred in the modern period for which we have instrumental weather records, and most have not been particularly large in a longer-term context (Ludlow et al. 2013; Sigl et al. 2015). Gaining an improved understanding of the impacts of explosive volcanism on climate and society is now of special concern, as a growing number of scientists and policymakers explore options for geoengineering to mitigate the temperature increases associated with the ongoing anthropogenic release of carbon dioxide and other greenhouse gases. Perhaps the most-discussed, and by some estimates most-affordable and technologically feasible, geoengineering scheme involves mimicking the cooling effects of explosive volcanism by artificially loading the stratosphere with sulphate particles (Moriyama et al. 2017). But there remain many unknowns, and much research is required to understand whether the deleterious impacts (listed comprehensively by Robock (2008)) of this form of geoengineering can be minimized, and if not, whether the benefits of controlling global average temperature rise may still outweigh the negative impacts from this form of geoengineering (Crutzen 2006; Lawrence and Crutzen 2017). Of particular concern in this respect are the likely impacts on regional patterns of precipitation and on the boreal summer monsoon, which may see diminished levels of rainfall. The fact that some 70% of world population depend upon the monsoon and related agriculture provides pause for thought in this respect (Mohtadi et al. 2016).

Studying the climatic impacts of past explosive volcanism provides one means to learn more about the potential impact of this form of geoengineering (Robock et al. 2013; Robock 2014). Natural archives offer a means of extending our history of explosive volcanism beyond the modern period, and the best archives at present comprise polar ice cores drilled in Greenland and Antarctica. These ice cores comprise a record of the annual deposition of snowfall compressed over time into layers of ice that are usually clearly visible (but dependent on many factors, including depth) and that can be counted backward from the present to identify each individual year with a high degree of accuracy. Efforts at ice-core-based reconstructions of past explosive volcanism began most prominently when the first long cores were recovered in 1966 in Greenland at Camp Century, and in Antarctica in 1968 at Byrd Station, and quickly began to supply unique details about the nature, timing and impacts of volcanic eruptions on regional to global scales through analysis of acid content in the ice, some of which represented the atmospheric fallout from large eruptions (Hammer 1977). By the 1990s, work had begun to progress from the study of the broader acid content of ice, as determined via electronical conductivity measurements, to specific measurements of the atmospheric deposition of (non-sea-salt) sulphate, a more direct product of volcanic activity. The volcanic record from the Greenland Ice Sheet Project 2 (GISP2) ice core represented a significant advance in this respect, with major explosive volcanic eruptions identifiable from large spikes in sulphate lying notably above average background deposition levels (Zielinski 1995; Zielinski et al. 1994, 1996). Such spikes are readily apparent in Fig. 3.3, which presents the non-sea-salt sulphate values measured in the GISP2 for Common Era, with each of the consecutive sulphate measurements taken between these dates representing the sulphate content of an approximately 2.5-year volume of ice (Mayewski et al. 1997), a sampling resolution that was state of the art at that time. Eruptions of the magnitude revealed by the GISP2 are capable of inducing

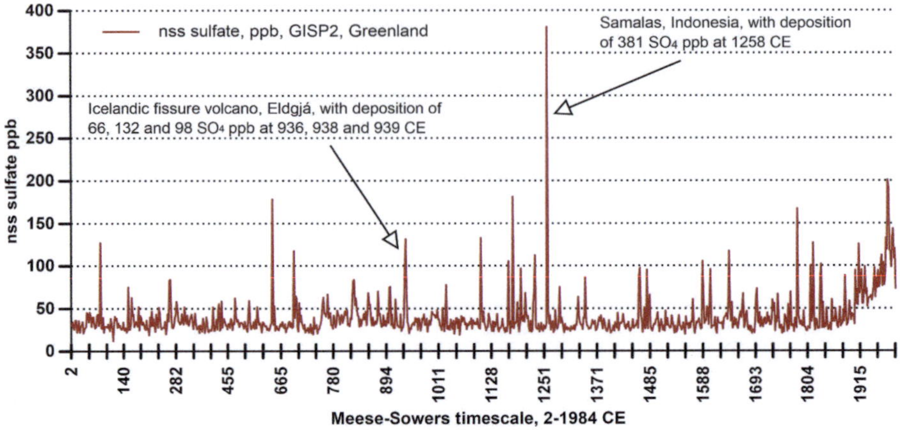

Fig. 3.3 Large "spikes" in non-sea-salt sulphate (SO_4) deposition above background levels in GISP2 ice-core samples (using the Meese-Sowers timescale) identify major explosive eruptions occurring prior to the modern period before which we have increasingly limited observational records of eruptions. The upward trend in background sulphate deposition visible during the twentieth century represents anthropogenic pollution sources

severe climatic disturbances and have often been associated with adverse human impacts recorded in both written records and archaeology from the ancient and medieval periods onward (e.g., Robock 2000; Cole-Dai 2010; Oppenheimer 2011; Kostick and Ludlow 2015; Ludlow and Manning 2016; Toohey et al. 2016; Guillet et al. 2017; Manning et al. 2017; Ludlow and Crampsie 2018).

To bring the written record of cold events from the Irish Annals to bear in examining the influence of explosive volcanism on the climate of the northeast Atlantic, Ludlow et al. (2013) employed the GISP2 ice-core record of volcanic sulphate deposition to derive a history of major volcanic eruptions over the broader period for which the Irish Annals provide coverage, 431 to 1649, identifying instances of notably elevated sulphate values to infer the occurrence of at least 48 major explosive volcanic eruptions (acknowledging that some volcanic signals could represent sulphate deposition from more than one eruption, if occuring closely together in time, particularly given ~2.5-year resolution of each sample).[6] Of the 65 historically credible cold events identified in the Irish Annals in addition to four further identified from other documentary sources,[7] some 37 (53.6%) occurred closely in

[6] A threshold of at least 55.16ppb SO_4 was applied, as per methods detailed in Ludlow et al. (2013), to identify ice-core SO_4 samples likely indicative of the occurrence of explosive volcanism.

[7] Seventy "cold events" were identified in total by Ludlow et al. (2013), but only 69 were included in their analysis of the correspondence of these events to volcanic signals identifiable from the GISP2 ice core, because of a gap in ice-core coverage (due to core damage) for a period encompassing the cold event dated to 586. Of the 70 cold events, 65 are identified from the evidence of the Irish Annals themselves, with the remainder identified from a range of other historical sources that supplement the Irish Annals as their coverage declines into the mid-seventeenth century. The section on the assessment of the reliability of documentary climatic data is thus specific to the

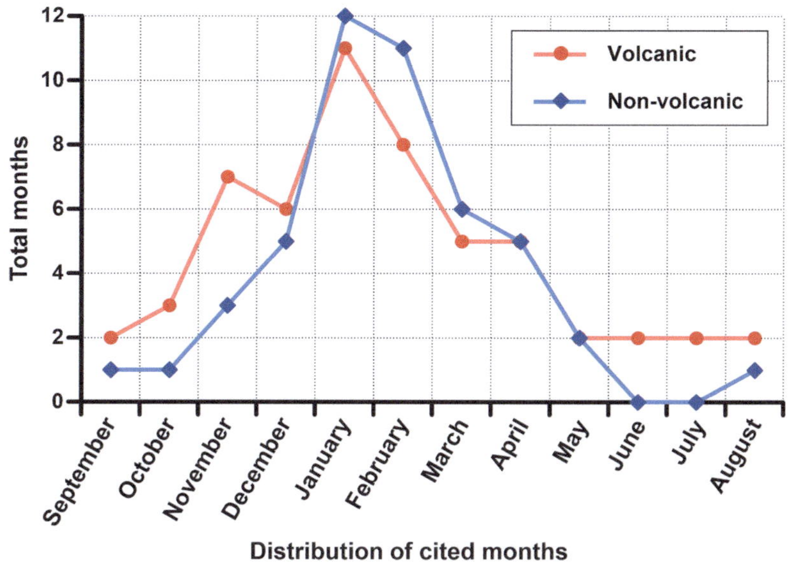

Fig. 3.4 Distribution of months reported in connection with severe cold events reconstructed for Ireland from the evidence of the Irish Annals and other documentary sources, 431–1649 (redrawn after Ludlow et al. 2013) extremes in the Irish Annals, 431–1649 CE. The *red line* (with *circles*) and *blue line* (with *diamonds*) represent, respectively, the months cited for cold events corresponding and not corresponding in time to GISP2 volcanic signals. Cold events associated with explosive volcanism are somewhat more evenly distributed throughout the year than cold events that appear unassociated, but winter months remain dominant in both

time (i.e., within 5 years) of GISP2 volcanic event signals. The likelihood that such a high degree of correspondence could occur purely by chance is less than 0.03% (a value based upon Monte Carlo randomization of the dates of volcanic event signals relative to the dates of cold events). This result strongly suggests the reality of explosive volcanism as a repeated cause of severe cold in the Irish region of the northeast Atlantic. An examination of the seasonality of the reported cold events also revealed a strong bias towards their occurrence in winter (Fig. 3.4). Here the Irish Annals are particularly significant in showing the impact of eruptions in a season rarely glimpsed in natural biological archives (or proxies) of climate that often primarily represent growing season climate (Hughes 2002). Thus, the historical evidence complements studies that have employed tree-ring-derived evidence to reveal the spring to summer season climatic impacts of historical explosive volcanism (e.g., Briffa et al. 1998; Bradley 2015; Sigl et al. 2015; Stoffel et al. 2015), thereby helping to provide a more complete assessment of the likely risks versus rewards of geoengineering via the stratospheric injection of sulphur dioxide, as well as provid-

evidence of the Irish Annals and the 65 events reliable events identified therefrom. See Ludlow et al. (2013) for full details.

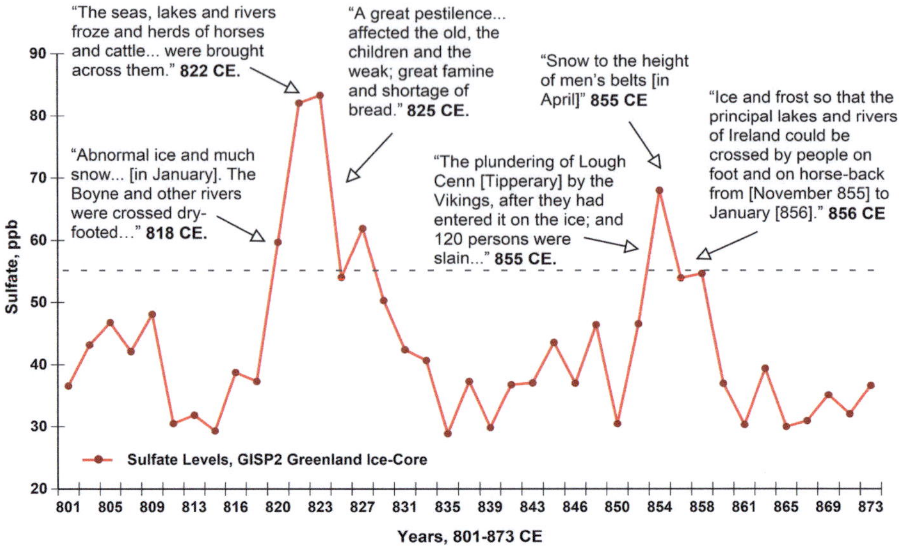

Fig. 3.5 GISP2 non-sea-salt (nss) sulphate values, 801–873, indicated with a *red dotted line*. The *horizontal dashed line* represents the sulphate deposition threshold employed by Ludlow et al. (2013) of 55.16 parts per billion (ppb) to identify potential volcanic signals. Values below, but near, this threshold, are also potentially indicative of volcanism, especially when closely timed with values above the threshold, in which case they may represent ongoing sulphate deposition from an earlier eruption. Reports of severe cold weather and major societal stresses from the Irish Annals that occur closely in time with these signals are also shown in summary (the descriptions have been abbreviated and paraphrased to fit the illustration; see Ludlow (2010) for full citations and source text attributions)

ing an aid in predicting (and hence any planning to mitigate) the likely climatic and related socioeconomic impacts of the next big eruption.

Figure 3.5 focuses on two closely spaced sequences of volcanic signals in the GISP2 that occur during the broadly neglected (from a historical climatological perspective) first millennium CE to further illustrate the results of the comparison between the GISP2 and Irish Annals record of cold, in addition to the variable societal impacts of volcanically induced severe cold. The figure shows the GISP2 ice-core sulphate measurements for the years 801–873, in which the years c.820, c.822, c.823 and c.827 exhibit elevated sulphate levels indicative of at least one and likely several volcanic eruptions (bearing in mind the uncertainty associated with the c.2.5 year resolution of each ice-core sample dated to these years). A further eruption is indicated in c.854, with high deposition continuing into the later 850s. The existence and broad dating of these volcanic signals can also be corroborated and refined by more recent and chronologically precise (i.e., higher-resolution) ice-core-based volcanic forcing data (Sigl et al. 2015) that identify substantial eruptions dated to 817 and 822 (both likely high-latitude Northern Hemispheric), and again in 853 and 859 (both likely tropical, with the 853 eruption being the more substantial). Closely occurring extremes of cold and societal stresses reported in the Irish Annals are also superimposed in Fig. 3.5. For 818 we can read of abnormal ice and snow (*Annals of Ulster and Annals of the Four Masters*), with "the materials for an oratory...brought from the lands of Connacht over Upper and Lower Loch Éirne..." in County Fermanagh (*Annals of Ulster*), such was the thickness of the ice, while for 822 we may read of freezing seas (most plausibly shallow-water inlets), rivers and lakes (*Chronicon Scotorum*, *Mageoghagan's Book*, *Annals of Ulster*), so that "herds of horses and cattle, and loads, were brought across them" (*Annals of Ulster*). Then, by 825, we may read of major societal stresses, namely, a "great pestilence that affected the old, the children and the weak" and a "great famine and shortage of bread" (*Annals of Ulster*, also *Chronicon Scotorum*). Severe cold is also documented in the aftermath of the GISP2 volcanic signal in c.854, including a report detailing "snow to the height of men's belts" in April of 855 (*Annals of Ulster*), with further cold in 856, such that "the chief lakes and rivers of Ireland could be crossed by people on foot and on horseback..." (*Chronicon Scotorum*, also *Annals of Ulster, Annals of the Four Masters and Fragmentary Annals*). Yet again, only a small number of years later (in 858), we read of "a rainy autumn destructive to the fruits of the earth" (*Annals of Ulster*) and "an autumn of famine" (*Fragmentary Annals*). These later societal stresses may well represent, at least in part, a cumulative impact of the preceding cold, but it is notable that in these specific cases, the Irish Annal do not explicitly cite any immediate societal stresses in connection with the years experiencing cold weather. This stands in contrast with many other instances of severe cold, such as the previously cited cold in 1115 with explicitly stated mass human and animal mortality in the same year, and can be taken as a caution to remember that medieval society could also prove robust to extreme weather. This additionally suggests that a fertile line of enquiry with relevance to contemporary concerns over societal resilience and adaptation will be the examination of the historical and socioeconomic context underlying instances in which extremes of apparently similar severity were *and were not* followed by major societal stresses.

The above material describes just one example from just one region, but more broadly illustrates how the combination of natural and written archives can provide insight into the climatic and human impacts of past volcanic eruptions as directly experienced and recorded by human witnesses, an approach that can be expanded to different time periods and regions. Such work is already ongoing. McCormick et al. (2007) have, for example, used the GISP2 data in combination with written evidence from early medieval Europe between 750 and 950 to examine the impact of explosive volcanism, while Büntgen et al. (2017) draw upon increasingly sophisticated (chronologically precise and accurate) ice-core data (based upon Sigl et al. 2015) as well as tree-ring data and written evidence to examine the climatic and societal impacts for Europe and China of the same volcanic events illustrated in Fig. 3.5, providing suggestive evidence that Icelandic volcano Katla may represent the source of the volcanic signals in c.822 and c.823. Still more remains to be done.

Science Informing History

Historians, archaeologists and other scholars of the human past are increasingly confronted with ever-more-detailed reconstructions of past climatic and environmental conditions from natural archives developed by their colleagues in the natural sciences (see Bradley (2015) for a review). These archives, and the palaeoclimatic and paleoenvironmental reconstructions that they allow, offer immense opportunities to scholars intent on understanding how human societies have interacted with their environments and coped with the hazards and extremes that accompany them. Yet these opportunities are tempered by considerable challenges in correctly interpreting what the available reconstructions can and cannot reveal of complex natural processes such as climate, with its part cyclic, part stochastic variation through space and time, as well as the potential limitations inherent in any given reconstruction from any given natural archive. Where scholars such as historians have chosen to engage with natural archives, it is not always clear that sufficient weight is given to the limitations of each natural archive or of the methods of (generally quantitative) analysis to which they are most suited.

Conversely, natural scientists are increasingly seeking lessons from the past in order to better understand the vulnerability of contemporary society to future climatic changes. Here they have availed of the human (archaeological and written) archives that are also now increasingly accessible. Many historical texts are, for example, translated and readily text-searchable in online digital platforms, and the texts of the majority of the Irish Annals are available on the *Corpus of Electronic Texts* (CELT) website.[8] But the very ease by which historical texts may now be consulted can allow an overly crude exploitation in work that pays too little attention to historical context, where nuances of meaning are overlooked and the reliability of texts and translations is afforded only rudimentary, if indeed any, consideration. A tendency also persists for studies in, or led by, the natural sciences to draw simplistic, crudely reductionist and deterministic links between reconstructed environmental changes, perceived societal responses and major historical events or transitions, in what can be described as "neo-environmental determinism" (Judkins et al. 2008). These concerns can only be resolved by embracing mixed quantitative-qualitative methods and by concerted multidisciplinary collaboration. Yet despite its oft-cited value, deeply interdisciplinary, multidisciplinary or transdisciplinary research and teaching concerning human-environmental relations of the type that may be characterized as a STEAM frontier still lacks fundamental institutional support (Roy et al. 2013; Adelman and Ludlow 2014). Efforts to overcome such barriers and pursue consilient approaches to the integration of historical and natural scientific evidence and methodologies are, however, ongoing, and the insights of scientific evidence regarding the human past are becoming increasingly hard to credibly ignore, even within the comfortable confines of specific disciplines such as history (McCormick 2011; Izdebski et al. 2016; Fuks et al. 2017; Newfield and Labuhn 2017; Haldon et al. 2018).

---

[8] See the Corpus of Electronic Texts (CELT) website, at www.ucc.ie/celt

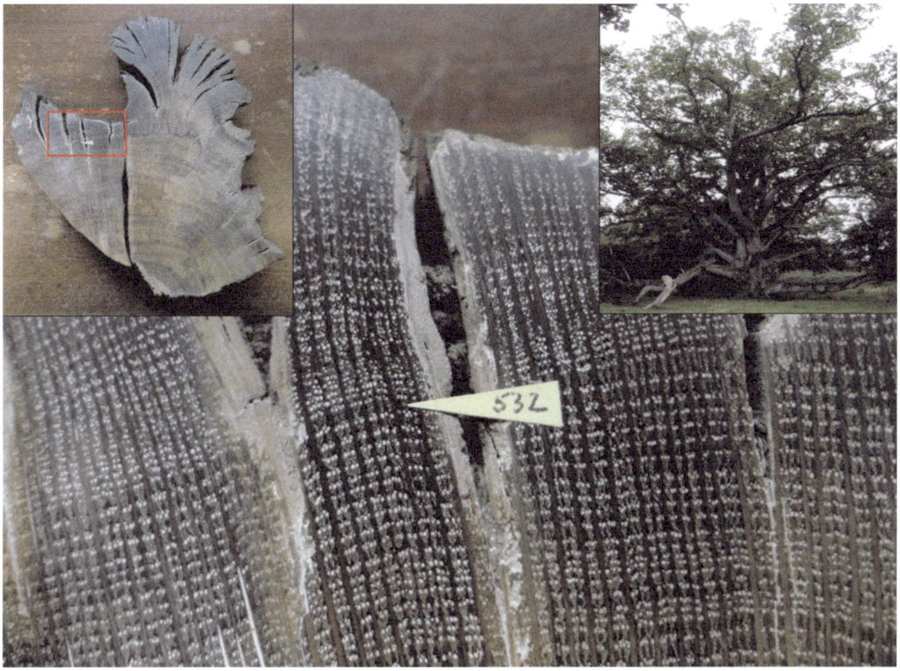

Fig. 3.6 Oak sample (Q9807A) from Deer Park townland, Antrim, Northern Ireland, with rings spanning the famous climate anomaly, c.536-550 CE. The wide-growth year 532 is marked, shortly after which (especially from 536 to 538) the rings are extremely narrow. We thank David Brown for permission to photograph this sample. The top left inset image shows the oak sample in full, while the top right inset shows a mature contemporary oak, known widely as the "King's Oak Tree", growing on the Charleville Forest Estate, Tullamore, Co. Offaly, Ireland

Of natural archives, Ireland benefits from the high-quality sessile and pedunculate oak tree-ring record developed in Queen's University Belfast starting from the 1970s by Michael G. L. Baillie, David Brown, Jonathan Pilcher and others. This record reveals yearly growing-season conditions for the past seven millennia with wide coverage across Ireland (Pilcher et al. 1984; Brown and Baillie 2012). Instances of unusually low growth have reliably identified years and decades of extreme weather, some clearly coinciding with historically recorded famine and mass human and animal mortality. Figure 3.6 presents a sample of an early medieval Irish oak preserved for approximately 1500 years in acidic waters of a bog in Ulster. The annual growth rings are clearly visible here, running up to the labelled year 532, shortly after which the tree begins to grow poorly, with narrow rings reflecting unfavourable growing conditions associated with a global climatic anomaly that began in 535 or 536 and continued until approximately 550, most likely the result of multiple closely occurring volcanic eruptions (Baillie 1994; Kostick and Ludlow 2015; Sigl et al. 2015; Toohey et al. 2016). This anomaly has been linked to famines and mortality in written sources from Ireland to China, with a report in the Irish

Annals for 538 (using McCarthy's (2005) chronological corrections) noting a "failure of bread" (*perdito panis*) (Baillie 1994, 1999; McCarthy 2008). This description evokes an image of notable agricultural difficulties that is all the more significant given the general rarity of events reported during this early period. The remainder of this section focuses upon the use of evidence from the Irish oaks as a complement to historical evidence from the Irish Annals in attempting to further examine whether meaningful environmental contexts exist for historical episodes of violence and conflict in medieval Ireland.

As much as they document extreme weather, the Irish Annals truly excel at the recording of violence and conflict, often waxing and waning in association with underlying political developments, throughout more than a millennium of contemporary recording. Recorded events range from the seizure of resources, scorched-earth tactics, forced migrations, slave raids, murders of societal elites, mass killings and battles, with added detail on locations, geographical spread and extent. Such intense coverage is likely to have arisen partly from the (at least nominal) incompatibility of violence with the scribes' Christian perspective and the direct impacts of this violence on their own safety and material wealth, partly from the culturally embedded and even ritualized conflict among Irish secular elites and rival families, territories and kingdoms, and partly from the dependence of Irish monasteries (and later the hereditary historians who maintained the Irish Annals from the thirteenth century onwards) on the patronage, and hence the military fortunes, of these elites (Fitzpatrick 1993; Aitchison 1998; Finney 1998; Geber 2012). The Irish Annals thus allow one to track and reconstruct the frequency, intensity, character and shifting spatial foci of violence and conflict during the long course of medieval Irish history in remarkable detail. The rich relevant information available can be shown by summarizing a selection of entries from just one year, 836, picked effectively at random from just one text, the *Annals of Ulster*. This reports that:

> The oratory of Kildare [monastery] was seized by Feidlimid [King of Munster] by force of arms from Forannán, abbot of Armagh". The monastery "was [later] plundered by the heathens [Norsemen] from Arklow [a Norse outpost, Leinster], and half of the church was burned". [The Norsemen next raided] "southern Brega [a north Leinster territory]...and they carried off many prisoners, and killed many and led away very many captives", then perpetrated "a cruel devastation of all the lands of Connacht...and inflicted slaughter in a battle won over the Déis Tuaisceirt [a north Munster people]."

The breadth and detail provided by the Irish Annals, combined with independent palaeoclimatic data such as that provided by the Irish oak ring-width record, thus offers the chance to examine associations between multiple conflict types, abrupt climatic changes, extreme weather events and other climatic stressors. The famous sixteenth-century woodcutting reproduced in Fig. 3.7 also illustrates one form of violence frequently found reported in the Irish Annals, namely cattle raiding, "a favourite recreation of the Gaelic ruling classes…commonly accompanied by the burning of houses and corn-ricks…" (Nicholls 1987, p. 414). This raiding represented an almost endemic and near ritualized form of violence in which young Irish nobility were to prove their prowess and amass wealth (Ó Corráin 2005), while perhaps also often serving wider political goals. Surprisingly, however, given that

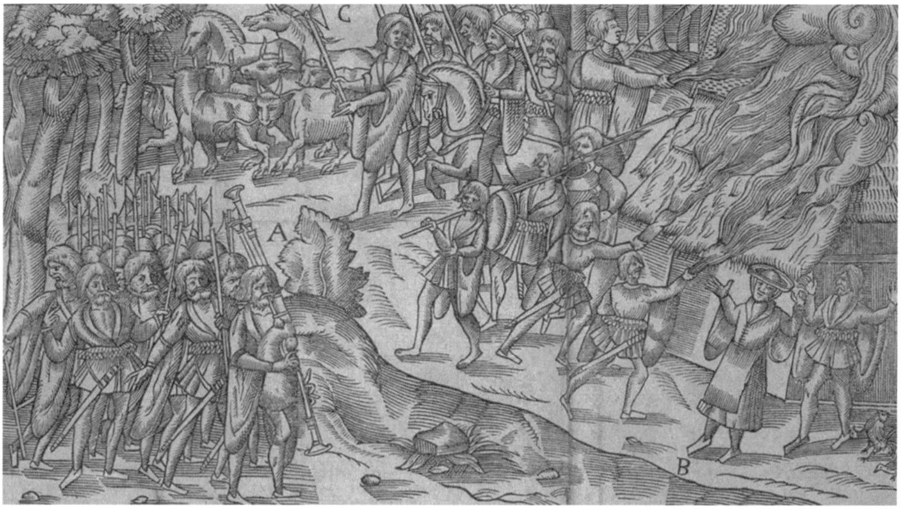

Fig. 3.7 Famous illustration of a cattle raid in Ireland from a sixteenth-century woodcut in John Derricke's *The Image of Irelande, with a Discoverie of Woodkarne* (1581). Image in the public domain (Wikimedia Commons). See Covington (2014) for commentary on the broader colonial context of this image

this behaviour focused upon the securing and accumulation of consumable resources, there has been little examination of whether climatic pressures played any systematic contributory role in promoting or modulating this activity.

One much-bemoaned characteristic that the Irish Annals share with annalistic sources elsewhere in Europe is a tendency to avoid explicit statements concerning the causality of recorded events (MacNeil 1919; White 1980). The texts do, however, provide some direct statements, including those linking weather conditions to violence and conflict, with the example highlighted below for the year 1465 classifiable as a case of violence promoted by scarcity-induced resource competition. Thus, for the year 1465, the *Annals of Connacht* describe "Exceeding great frost and snow and stormy weather this year, so that no herb grew in the ground and no leaf budded on a tree until the feast of St. Brendan [May 16], but a man, if he were the stronger, would forcibly carry away the food from the priest in church…". Fitzpatrick (1993) and Ó Corráin (2005) note further examples, including instances of internal migration and population displacement within Ireland with suggestive links to pressures from extreme weather and/or famine, and instances in which the texts suggest some implicit causal or contributory association (e.g,. between harsh weather and the raiding of churches as detailed in 1077) by reporting both events in the same entry and sentence, even if not explicitly acknowledging a connection.

Detailed paleoclimatic data is essential in attempting to systematically identify linkages between extreme weather, violence and conflict, because weather extremes of different character (i.e., affecting certain seasons, occurring rapidly or gradually) may plausibly promote violence and conflict that varies distinctly in type, intensity, duration and extent (Zhang et al. 2010; Bai and Kung 2011; Hendrix and Salehyan 2012; Ludlow and Crampsie 2018). The abundant seasonally specific recording in the Irish Annals of extreme precipitation (surplus, deficit), temperatures (heat, cold) and windiness (windstorms, windy seasons) and associated impacts (population movements, scarcity and famine, human and animal disease and mass mortality) can be employed to examine the pathways and societal dynamics underlying any statistically observed climate-conflict associations. But despite the density of reporting in the Irish Annals, its record of extreme weather events is not fully complete, and the reporting of weather was not the sole purpose of the Irish annalistic record (Ludlow 2012). The information we may glean from the Irish Annals can therefore be usefully augmented by data from natural archives, as illustrated in the example below.

Amongst other factors, the growth of Irish oaks is influenced in an interactive manner by an array of weather conditions, including temperatures and precipitation (via soil moisture levels). However, they are generally most sensitive in Ireland to spring and summer soil moisture (García-Suárez et al. 2009; Cook et al. 2015). Figure 3.8 shows the average annual growth of oaks across Ireland from 728 to 748, a period during which the trees register a profound decrease in growth for the years 737 and 738. This decrease is of a scale only rarely matched in the remainder of the first millennium. Despite the magnitude of the drought that can be inferred from this growth decrease, and the fact that the Irish Annals are not infrequent recorders of drought (Ludlow 2010), there is no mention of weather conditions in these specific years. We can, however, find written corroboration of the Irish oaks in the form of a description from Britain of "A great drought [that] rendered the land infertile" for the year 737 as reported by a continuator of Bede's *Ecclesiastical History* (Colgrave and Mynors 1969, p. 573). Meteorological conditions leading to severe drought in Britain and Ireland often involve persistent anticyclonic circulation regimes that are typically spatially extensive (Dukes and Eden 1997; Sweeney 1997; Noone et al. 2017), and the report from Bede's continuator can thus be deemed fairly representative of the meteorological circumstances prevailing in Ireland at this time. Despite the omission of this drought in the Irish Annals, the texts are not in fact silent for these years, and do record events of other kinds, including violence and conflict. In this respect, a straightforward count of deaths of named (and hence elite) individuals in violent conflict from 728 to 748 in the *Annals of Ulster* reveals both a persistent low level of reporting of this form of violence, accompanied by a striking elevation of such deaths reported for 738 (Fig. 3.8), conspicuously matching the second consecutive spring-summer of severe drought identifiable through the independent evidence of the Irish oaks.

How should this result be interpreted, then? Certainly, it is consistent with the above-cited evidence of linkages between weather-related scarcity and increased (including violent) competition over these scarcened resources. It is also suggestive of how independent evidence from natural archives can be brought to bear in interpreting patterns of reporting in the Irish Annals. But there are important and interrelated questions that cannot be answered by a pointing to a single coincidence in timing between inferred drought and elevated violence. Is, for example, this coinci-

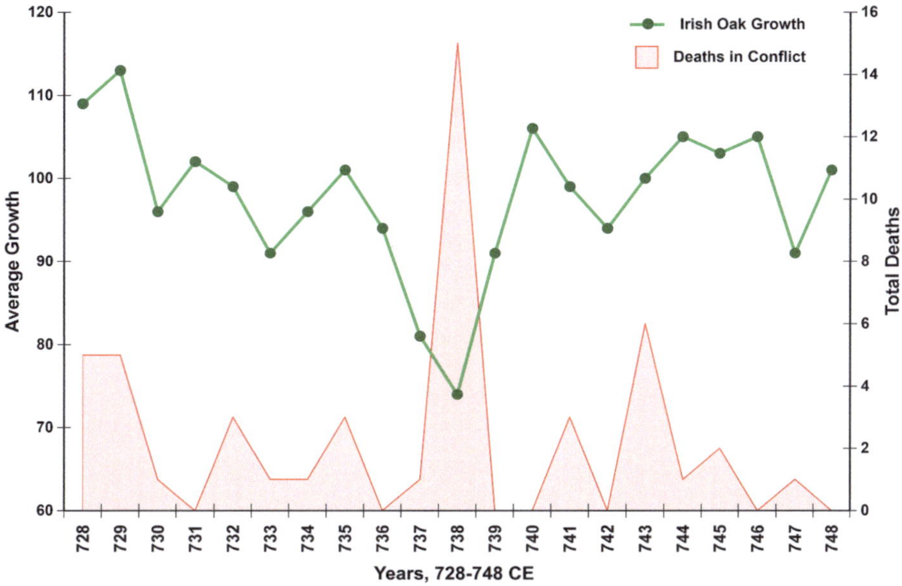

Fig. 3.8 Indexed Irish oak ring width growth data, with a count of deaths in conflict from the *Annals of Ulster*, 728–748. Irish oak data courtesy of Mike Baillie and David Brown, varying around a long-term average index value of 100. Severe drought conditions are implied by dramatically reduced growth in the years 737 and 738, with a notable elevated reporting of violent deaths in conflict in 738

dence in timing purely random (statistically, such coincidences can be expected to occur on a certain number of occasions), and if it is not random in this specific case (accepting for the purposes of discussion that the drought did indeed meaningfully contribute to elevated violence in 738), then we might next ask whether droughts (or other extreme weather conditions) *systematically* correspond with heightened violence and conflict, and if so, whether and how the strength of that correspondence varies through time and space? To begin answering such questions, there is a clear need to conduct statistical comparisons over multiple drought events, and to closely examine the underlying historical socioeconomic, geographical, cultural and political contexts that will facilitate and mediate any links between extremes and any given social outcome such as violence and conflict. The combination of natural and human archives available to scholars of the human past provides a means of answering such questions, and this endeavour will guide our assessment of how past experiences may be relevant or informative for contemporary society.

In what ways, then, can historians begin to integrate evidence from natural archives and understandings of the potential for linkages between climate and society into their historical narratives? Returning to the year 738, it is clear that many of these deaths are of social elites occurring in connection with the Battle of Áth Senaig (Ballyshannon in Co. Kildare), a "great battle" in which "so many fell... that we find no comparable slaughter...throughout all preceding ages" (*Annals of Ulster*).

Even allowing for exaggeration in its description, with evidence of more than a passing scribal interest in this battle highlighted by the description's embellished phraseology and brief transition to the use of the present tense, this appears to be a historically credible and important event that is found in multiple Irish annalistic texts from different eras, including the fourteenth century *Annals of Tigernach* and seventeenth century *Annals of the Four Masters* (note that this latter text reports the event incorrectly under 733 before applying the chronological corrections of McCarthy (2005)). The battle itself forms part of the iterative conflict between the Uí Néill and the Laigin (two powerful and not infrequently warring septs in early medieval Ireland) which is eventually won by the Uí Néill, contributing to a rewriting of the political landscape of early medieval Ireland (Ó Cróinín 1995). Historians may ask whether it is incidental that this battle took place in the second (and not the first) year of consecutive severe drought, and whether agricultural difficulties may have played upon any existing tensions and fault lines discernable from a close reading of the available texts. Ireland in this period was certainly not lacking in such tensions (see, e.g., Maney 2000–2001). It might also be asked whether the drought disadvantaged one side more than the other, with the specific timing of the battle thus representing an attempt to exploit one side's environmentally mediated vulnerabilities. If future statistical analyses reveal such coincidences between drought and conflict to be frequent in medieval Ireland, then our historical writings can only be enriched when armed with the evidence of natural archives and with an understanding of the influences of climate (and other environmental forces) on human history. Historians, furthermore, need not only be "end users" of such evidence and insights, but can and must play a fundamental collaborative role in ensuring that the historical timeseries of violence and conflict that any such statistical analyses will inevitably be based upon are generated with a nuanced understanding of the historical sources and periods from which they are drawn.

In understanding the potential influences of climate, we must stress that the evidence of the Irish Annals also makes it clear that medieval Irish society was not merely a passive victim of extreme weather. Responses and coping strategies were clearly available to mitigate the impacts of extreme weather and constrain any related violence and conflict. For the year 1050, the *Annals of the Four Masters* thus describe how:

> Much inclement weather happened in the land of Ireland, which carried away corn, milk, fruit, and fish, from the people, so that there grew up dishonesty among all, [so] that no protection was extended to church or fortress, gossipred or mutual oath, until the clergy and laity of Munster assembled, with their chieftains, under Donnchadh, son of Brian, i.e. the son of the King of Ireland, at Cill-Dalua [Killaloe, Co. Clare], where they enacted a law and a restraint upon every injustice, from small to great. God gave peace and favourable weather in consequence of this law.

In this entry we see a further example of scarcity-induced resource competition promoting negative social behaviours, but also secular and ecclesiastical elites acting in concert to restore order. This entry also strongly suggests the ways in which extreme weather might play into the politics of the era. Subtext can be read in the description on many levels, raising questions of authorship, intended readership, purpose and even its historicity (see discussion in Baker et al. (2017)). We can note in this respect that while the adverse weather and behaviours appear widespread throughout Ireland, it is apparently only the clergy and laity of Munster (and not rival provinces) that resolve to act, thereby demonstrating their moral leadership and claiming divine sanction in an era when the high kingship of Ireland was contested between different provincial powers. Donnchadh holds the meeting at Killaloe, moreover, near the capital of his late father Brian Boru (i.e., the "Brian" referenced in the description), perhaps the only Irish provincial king to have successfully controlled all of Ireland as High King. Donnchadh's alleged success in this auspicious location would magnify his credibility in following his father in attempting to gain the ever-contested High Kingship of Ireland. It is noteworthy, therefore, that most other Irish annalistic texts excepting the Munster-based *Annals of Inisfallen* make no mention of Donnchadh's kingly success this year (e.g., the *Annals of Ulster* incorporating at this time much annalistic material written in Armagh monastery, traditionally patronized by Uí Néill dynasty who were rivals to Munster princes such as Donnchadh (McCarthy 2008)).

Natural archives again provide a valuable contribution to understanding the context in which this entry was written and whether indeed it has any basis in historical reality. Ireland is not the only European region to benefit from abundant oak ring width chronologies. These are in fact widely available across temperate Europe, and because the ring-width growth of these oaks responds clearly (if not solely) to spring and summer soil moisture availability, various hydroclimatic reconstructions have been made from their evidence, including the recent Old World Drought Atlas (OWDA) that provides coverage for much of Europe (Cook et al. 2015). More formally, the OWDA is a reconstruction of the self-calibrating Palmer Drought Severity Index (PDSI, reflecting spring-summer soil moisture) using a network of 106 precipitation-sensitive tree-ring-width chronologies and offers historians and other scholars the ability to identify years and regions of surplus and deficit soil moisture for the past two millennia. Figure 3.9 presents a map of reconstructed PDSI values across Europe, in which browner equates to less soil moisture and bluer more. The year mapped is 1050, in which the PDSI values are starkly negative across much of Europe, indicating severe drought conditions in which Ireland appears as a heavily impacted region. The evidence of the European oaks hence corroborates the reporting from the *Annals of the Four Masters*, at least regarding the occurrence of the unspecific "inclement weather", going further by identifying the involvement of drought, and ultimately allowing us to make a more informed interpretation of the text and the events it reports.

What this case suggests, then, is a medieval society capable of rapidly enacting measures to restore stability during severe weather and related food scarcity, but conducted (or at least reported) in a manner that is revealing of the ways in which environmental forces intersected with the political realities and ambitions of the period (Baker et al. 2017). The events of 1050 do not constitute the only instance of coping responses and mitigation measures detailed in the Irish Annals (e.g., see Ó Corráin (2005)), and further study will shed light on their evolution and efficacy, which may have varied considerably through time, waxing and waning alongside

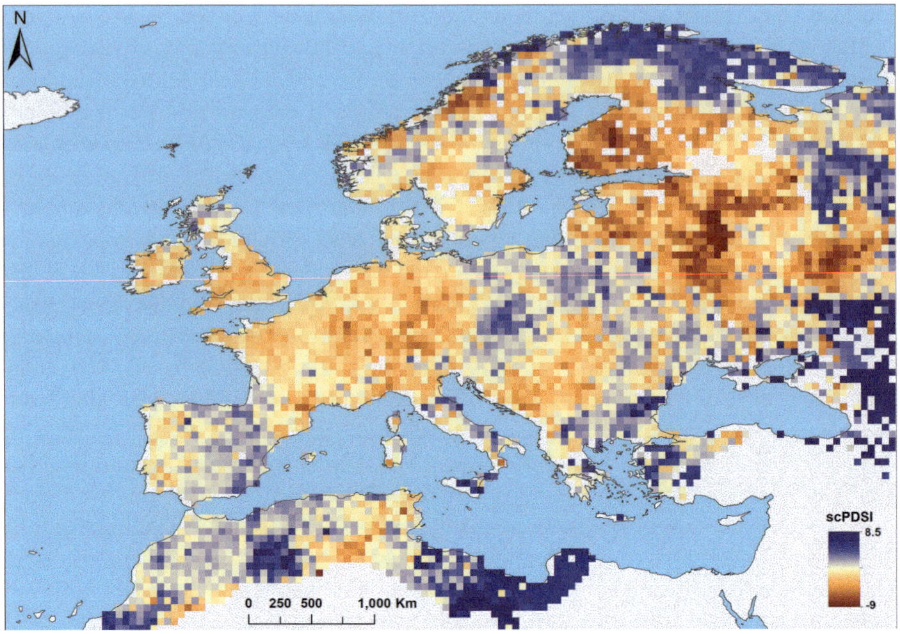

Fig. 3.9 Reconstructed self-calibrating Palmer Drought Severity Index (scPDSI) values for the year 1050, showing a widespread severe spring-summer soil moisture deficit. The map is drawn using data from Cook et al. (2015))

the influence of the church or the willingness of ecclesiastical and secular powers to cooperate, for example. If so, this may ultimately be discernable as periods that exhibit a more or less marked increase in violence and conflict during extreme weather and subsistence crises. The likelihood that a sizeable fraction of medieval Irish society will have believed extreme weather and subsistence crises to be divinely wrought (e.g., as punishment) and hence addressable by penitential behaviour also holds interesting implications regarding the degree to which other potentially more concrete mitigation or coping mechanisms were likely to be pursued, and may have even diminished efforts to achieve a fuller adaptation to extreme weather and prevent the onset of subsistence crises to begin with.

This section has sought to highlight the potential of evidence from natural archives to complement evidence from written sources in examining human relationships with the environment. Capitalizing upon the potential of combining these sources of evidence will allow us to answer increasingly complex questions, not least whether and how different types and severities of extreme weather, occurring in different seasons, with different regularity or against differing climatic backgrounds provoked meaningfully different societal impacts and responses. The question has often been posed, for instance, of whether the sudden occurrence of extreme cold weather will have a greater impact if it occurs in a generally colder versus warmer period? Answering this involves deceptively complex considerations of short-term weather variations occurring against a background of longer-term climatic changes, as well as considerations of societal adaptaton and preparedness. Has the society effectively adapted to generally colder conditions, such that it has achieved greater resilience to the ocurrence of extreme conditions? Or will the onset of extreme cold have a greater impact in a society already stressed by the effort to adapt to (or even failure to adapt to) more generally colder conditions? In a generally warmer period, will the onset of extreme cold weather be more impactful for a lack of preparedness and adaptation to cold, or will adaptation to the more generally warmer and potentially more agriculturally productive conditions allow society to better sustain itself even through the surprise visitation of extreme cold? Variables such as technology (military and agricultural), population size and mix, and evolving political, economic and belief systems, can all also be expected to mediate the impacts of extreme weather on society. These questions are complex, but can be best answered by STEAM approaches that combine the methods, insights and evidence bases of the sciences, arts and humanities.

Conclusion

Historical climatology and climate history are inherently interdisciplinary fields that illustrate the efficacy of STEAM approaches, sitting at the intersection between environmental history and climatology (Pfister 2007, 2010). Historical climatology as a field has a long pedigree, with recognizably relevant works of scholarship published as early as the 1920s and antecedents extending into the eighteenth century. Only in the most recent decades, however, has it achieved a significant prominence and can now be clearly defined as an interdisciplinary STEAM frontier. This chapter, recognizing the growing awareness in the twenty-first century of the potential associations between climate change, extreme weather and social and political conflict, has discussed the potential and challenges of engaging in historical climatology and climate history, drawing upon case studies illustrative of the two main interrelated strands of work in these fields. Adopting didactic framing, the first strand can be usefully charaterized as a case of history informing science, taking the example of written evidence of the climatic impact of explosive volcanism preserved within the Irish Annals. The second can be seen as a case of science informing history, taking the example of tree-ring-based evidence of past extreme weather informing studies of historical societal vulnerability to sudden environmental hazards.

Perhaps unsurprisingly, given increasing concern over the likely human impacts of global warming, many natural scientists are taking their paleoclimatic reconstructions and cross-referencing the patterns they see therein against human history. Often this work exhibits little hesitation in making simple causal connections between climatic trends, extreme weather and major phases or events in human history, be they the rise and fall of empires or trends in the occurrence of war and conflict. Whether such studies always give sufficient consideration to the complexities of how humans interact with their environment, and to the role of human agency, is debatable (see the comments of Fan (2010), for example). Historians and other scholars can collaborate, and are perhaps even obligated to collaborate, with their colleagues in the natural sciences in order to overcome tendencies towards environmental determinism and reductionism, and more credibly inform current debates. Undertaking such collaboration will mean, however, that historians must engage open-mindedly about the potential role of climate in human history. Campbell (2010, p.310) has thus remarked that:

> …with a mounting body of scientific evidence to draw upon, there has never been a better opportunity to explore the interrelationships between past environmental events and processes and the course of social and economic change. The time has surely come to acknowledge that—alongside the class struggle, invisible hand ofthe market, creation and diffusion of technology and knowledge, and an array of human institutions (including many intended to mitigate and counteract the risks arising from environmental hazards)—'Nature' was an historical protagonist in its own right.

If we are to successfully pursue STEAM approaches, the effort will involve surmounting disciplinary conservatism and mistrust and learning to translate between different disciplinary "languages". Challenges include scientists' distrust of what has been called "soft" data and climate modelling periodicity (fitting the period to the model rather than vice versa). It will mean developing (and accepting the legitimacy of) hybrid methodologies and research goals (not "forcing" the application of scientific methodologies, or arts and humanities interventions, but facilitating collaboration where applicable). It will also need real, not just nominal, institutional support. Whether many historians will begin to pursue this avenue of research (within climate history or beyond) will largely depend on whether they, and their colleagues and institutions, feel that this qualifies as "doing" history. For us it absolutely does, whilst for others it probably doesn't. But if historians really wish to explore all the avenues to which our expertise can be applied, then we have choices to make in how we define our discipline and even in the scope of our hiring decisions. Against this background, on 28 November 2012, the *Initiative for the Science of the Human Past at Harvard* convened a special meeting of leading scientists in ice-core palaeoclimatic records, dendrochronology and dendroclimatology, with scholars expert in economic and environmental history and historical climatology (Ludlow 2013). The evidence of high-resolution tree-ring and ice-core archives was considered here in tandem with written records to identify climatic and societal crises particularly illustrative of the challenges and potential of unifying these disparate evidence bases. This and the work of other initiatives such as the PAGES Volcanic Impacts on Climate and Society (VICS) Working Group, the Yale Climate and History Initiative, the Princeton Climate Change and History Research Initiative, and the PAGES Climate Reconstruction and Impacts from the Archives of Societies (CRIAS) Working Group can perhaps be seen as signs of the evolution of a STEAM historical climate studies approach, which will have critical importance in helping us address the challenges of global warming, extreme weather and the attendant social-political crises that are already occurring and will continue to if effective mitigation and adaption strategies are not pursued with urgency.

References

Adano WR, Dietz T, Witsenburg K, Zaal F (2012) Climate change, violent conflict and local institutions in Kenya's drylands. J Peace Res 49(1):65–80

Adelman J, Ludlow F (2014) The past, present and future of environmental history in Ireland. Proc R Ir Acad C Archaeol Celtic Stud Hist Linguist Lit 114:359–391

Aitchison NB (1998) Regicide in early medieval Ireland. In: Halsall G (ed) Violence and society in the early medieval West. The Boydell Press, Woodbridge, pp 108–125

Bai Y, Kung JK (2011) Climate shocks and Sino-nomadic conflict. Rev Econ Stat 93(3):970–981

Baillie MGL (1994) Dendrochronology raises questions about the nature of the AD 536 dust-veil event. Holocene 4(2):212–217

Baillie MGL (1999) Exodus to Arthur: catastrophic encounters with comets. B.T. Batsford, London

Baker L, Brock S, Cortesi L, Eren A, Hebdon C, Ludlow F, Stoike J, Dove M (2017) Mainstreaming morality: an examination of moral ecologies as a form of resistance. J Study Relig Nat Cult 11(1):23

Barnett J (2003) Security and climate change. Glob Environ Chang 13:7–17

Barnett J, Adger WN (2007) Climate change, human security and violent conflict. Polit Geogr 26(6):639–655

Bell WT, Ogilvie AEG (1978) Weather compilations as a source of data for the reconstruction of European climate during medieval period. Clim Chang 1:331–348

Bradley RS (2015) Paleoclimatology: reconstructing climates of the quaternary, 3rd edn. Academic, Oxford

Brázdil R, Pfister C, Wanner H, Von Storch H, Luterbacher J (2005) Historical climatology in Europe – the state of the art. Clim Chang 70:363–430

Briffa K, Jones P, Schweingruber F, Osborn T (1998) Influence of volcanic eruptions on northern hemisphere summer temperature over the past 600 years. Nature 393:450–455

Britton CE (1937) A meteorological chronology to A.D. 1450. Geophys Mem 8:3–177

Brooks CEP (1928) Historical climatology of England and Wales. Q J R Meteorol Soc 54:309–317

Brown DM, Baillie MGL (2012) Confirming the existence of gaps and depletions in the Irish oak tree-ring record. Dendrochronologia 30(2):85–91

Buhaug H, Nordkvelle J, Bernauer T et al (2014) One effect to rule them all? A comment on climate and conflict. Clim Chang 127(3–4):391–397

Campbell BMS (2010) Nature as historical protagonist: environment and society in pre-industrial England. Econ Hist Rev 63(2):281–314

Cole-Dai J (2010) Volcanoes and climate. Wiley Interdiscip Rev Clim Chang 1(6):824–839

Colgrave B, Mynors RAB (eds) (1969) Bede's ecclesiastical history of the English people. Clarendon Press, Oxford

Cook ER et al (2015) Old world megadroughts and pluvials during the common era. Sci Adv 1:e1500561

Covington S (2014) 'Realms so barbarous and cruell': Writing Violence in Early Modern Ireland and England. History 99(336):487–504

Crutzen PJ (2006) Albedo enhancement by stratospheric sulfur injections: a contribution to resolve a policy dilemma? Clim Chang 77(3–4):211–219

De Boer JZ, Sanders DT (2002) Volcanoes in human history: the far-reaching effects of major eruptions. Princeton University Press, Princeton

De Châtel F (2014) The role of drought and climate change in the Syrian uprising: untangling the triggers of the revolution. Middle East Stud 50:521–535

Dukes M, Eden P (1997) 'Phew! What a scorcher': weather records and extremes. In: Hulme M, Barrow E (eds) Climates of the British Isles: present, past and future. Routledge, London/New York, pp 262–295

Dunphy G (2010) Annals. In: Dunphy G (ed) The encyclopedia of the medieval chronicle, vol 1. Brill, Leiden, pp 45–52

Fan K-w (2010) Climatic change and dynastic cycles in Chinese history: a review essay. Clim Change 101(3–4):565–573

Finney J (1998) Warfare in late pagan and early Christian Ireland c.500AD to c.750. Mediev Life 9:3–10

Fischer E et al (2007) European climate response to tropical volcanic eruptions over the last half millennium. Geophys Res Lett 34:L05707

Fitzpatrick E (1993) Raiding and warring in monastic Ireland. Hist Irel 1(3):13–18

Fjelde H, von Uexkull N (2012) Climate triggers: rainfall anomalies, vulnerability and communal conflict in sub-Saharan Africa. Polit Geogr 31(7):444–453

Fuks D, Ackermann O, Ayalon A, Bar-Matthews M, Bar-Oz G, Levi Y, Maeir AM, Weiss E, Zilberman T, Safrai Z (2017) Dust clouds, climate change and coins: consiliences of palaeoclimate and economy in the Late Antique southern Levant. Levant 49(2):205–223

Gao C, Ludlow F, Amir O, Kostick C (2016) Reconciling multiple ice-core volcanic histories: the potential of tree-ring and documentary evidence, 670–730 CE. Quat Int 394:180–193

García-Suárez AM, Butler CJ, Baillie MGL (2009) Climate signal in tree-ring chronologies in a temperate climate: a multi-species approach. Dendrochronologia 27:183–198

Geber J (2012) Comparative study of Perimortem weapon trauma in two early medieval skeletal populations (AD 400–1200) from Ireland. Int J Osteoarchaeol. https://doi.org/10.1002/oa.2281

Graham E (2004) The emerald isle turns white. Snow and very low surface temperatures over Ireland during Christmas 2000. Weather 59(1):15–19

Greetham D (1992) Textual scholarship: an introduction. Routledge, London

Guillet S, Corona C, Stoffel M, Khodri M, Lavigne F, Ortega P, Eckert N, Sielenou PD, Daux V, Churakova Sidorova OV, Davi N, Edouard J-L, Zhang Y, Luckman BH, Myglan VS, Guiot J, Beniston M, Masson-Delmotte V, Oppenheimer C (2017) Climate response to the Samalas volcanic eruption in 1257 revealed by proxy records. Nat Geosci 10:123–128

Haldon J, Lee M, Newfield TP, Chase AF, Izdebski A, Guzowski P, Labuhn I, Roberts N (2018) History meets palaeoscience: consilience and collaboration in studying past societal responses to environmental change. Proc Natl Acad Sci 115(13):3210–3218

Hammer CU (1977) Past volcanism revealed by Greenland ice sheet impurities. Nature 270:482–486

Hassan F (2000) Environmental perception and human responses in history and prehistory. In: McIntosh RJ, Tainter JA, McIntosh SK (eds) The way the wind blows: climate, history, and human action. Columbia University Press, New York, pp 121–140

White H (1980) The value of narrativity in the representation of reality. Crit Inq 7(1):5–27

Hendrix CS, Salehyan I (2012) Climate change, rainfall, and social conflict in Africa. J Peace Res 49(1):35–50

Hickey K (2011) The historic record of cold spells in Ireland. Ir Geogr 44(2–3):303–321

Hsiang SM, Burke M (2014) Climate, conflict and social stability: what does the evidence say? Clim Chang 123(1):39–55

Hsiang SM, Burke M, Miguel E (2014) Reconciling climate-conflict meta-analyses: reply to Buhaug et al. Clim Chang 127(3–4):399–405

Hughes MK (2002) Dendrochronology in climatology – the state of the art. Dendrochronologia 20:95–116

Izdebski A, Holmgren K, Weiberg E, Stocker SR, Büntgen U, Florenzano A, Gogou A, Leroy SAG, Luterbacher J, Martrat B, Masi A, Mercuri AM, Montagna P, Sadori L, Schneider A, Sicre M-A, Triantaphyllou M, Xoplaki E (2016) Realising consilience: how better communication between archaeologists, historians and natural scientists can transform the study of past climate change in the Mediterranean. Quat Sci Rev 136:5–22

Izdebski A, Pickett J, Roberts N, Waliszewski T (2016) The environmental, archaeological and historical evidence for regional climatic changes and their societal impacts in the eastern Mediterranean in late antiquity. Quat Sci Rev 136:189–208

Jones PD, Briffa KR, Osborn TJ, Lough JM, van Ommen TD, Vinther BM, Luterbacher J, Wahl ER, Zwiers FW, Mann ME, Schmidt GA, Ammann CM, Buckley BM, Cobb KM, Esper J, Goosse H, Graham N, Jansen E, Kiefer T, Kull C, Küttel M, Mosley-Thompson E, Overpeck JY, Riedwyl N, Schulz M, Tudhope AW, Villalba R, Wanner H, Wolff E, Xoplaki E (2009) High-resolution palaeoclimatology of the last millennium: a review of current status and future prospects. Holocene 19:3–49

Judkins G et al (2008) Determinism within human-environment research and the rediscovery of environmental causation. Geogr J 174:17–29

Kelley CP, Mohtadi S, Cane MA, Seager R, Kushnir Y (2015) Climate change in the fertile crescent and implications of the recent Syrian drought. Proc Natl Acad Sci 112:3241–3246

Kostick C, Ludlow F (2015) The dating of volcanic events and their impacts upon European climate and society, 400–800 CE. Eur J Post-Class Archaeol 5:7–30

Kravitz B, Robock R (2011) Climate effects of high-latitude volcanic eruptions: role of the time of year. J Geophys Res Atmos 116:D01105. https://doi.org/10.1029/2010JD014448

Lamb HH (1995) Climate, history and the modern world, 2nd edn. Routledge, London

Lawrence MG, Crutzen PJ (2017) Was breaking the taboo on research on climate engineering via albedo modification a moral hazard, or a moral imperative? Earth's Future 5(2):136–143

Le Roy Ladurie E (1971) Times of feast, times of famine: a history of climate since the year 1000. George Allen and Unwin Ltd, London

Ludlow F (2010) The utility of the Irish annals as a source for the reconstruction of climate. Unpublished PhD thesis, School of Natural Sciences, Trinity College Dublin, Dublin

Ludlow F (2012) Assessing non-climatic influences on the record of extreme weather events in the Irish annals. In: Duffy PJ, Nolan W (eds) At the anvil: essays in honour of William J. Smyth. Geography Publications, Dublin, pp 93–133

Ludlow F (2013) Workshop examines climate change and human response in the history of western Eurasia from AD 1–1600. Harvard Univ Cent Environ Newsl 5(1):29 Available at: environment.harvard.edu/node/3270

Ludlow F (2017) Volcanology: chronicling a medieval eruption. Nat Geosci 10(2):77–78

Ludlow F, Crampsie A (2018) Environmental history of Ireland, 1550–1730. In: Ohlmeyer J (ed) Cambridge history of Ireland: volume 2, early modern Ireland, 1550–1730. Cambridge University Press, Cambridge, pp 671–704

Ludlow F, Manning JG (2016) Revolts under the Ptolemies: A paleoclimatic perspective. In: Collins JJ, Manning JG (eds) Revolt and resistance in the ancient classical world and the near East: the crucible of empire, Culture and History of the Ancient Near East Series. Brill, Leiden, pp 154–171

Ludlow F, Stine AR, Leahy P, Murphy E, Mayewski PA, Taylor D, Killen J, Baillie MGL, Hennessy M, Kiely G (2013) Medieval Irish chronicles reveal persistent volcanic forcing of severe winter cold events, 431–1649 CE. Environ Res Lett 8:024035. https://doi.org/10.1088/1748-9326/8/2/024035

MacNeil H (2000) Trusting records: legal, historical and diplomatic perspectives. Kluwer Academic Publishers, Dordrecht/London/Boston

MacNeil E (1919) Phases of Irish history. M. H. Gill & Son, Dublin

Maney L (2000–2001) Rebuke and revision in the early Irish annals: the death-notices of Muirchertach mac Ercae [†534] and Finsnechta mac Cellaig [†808]. In: Proceedings of the Harvard Celtic Colloquium, 20–21, pp 64–85

Manning JG, Ludlow F, Stine AR, Boos W, Sigl M, Marlon J (2017) Volcanic suppression of Nile summer flooding triggers revolt and constrains interstate conflict in ancient Egypt. Nat Commun 8. https://doi.org/10.1038/s41467-017-00957-y

Mayewski PA et al (1997) Major features and forcing of high-latitude northern hemisphere atmospheric circulation using a 110,000-year-long glaciochemical series. J Geophys Res 102(C12):26,345–26,366

McCarthy D (1998) The chronology of the Irish annals. Proc R Ir Acad 98C:203–255

McCarthy D (2001) The chronology and sources of the early Irish annals. Early Mediev Europe 10(3):323–341
McCarthy D (2005) Irish chronicles and their chronology, 4th edn. Available: www.cs.tcd.ie/Dan.McCarthy/chronology/synchronisms/annals-chron.htm. Last updated: 21/03/05. Accessed 1 Nov 2012
McCarthy D (2008) The Irish annals: their genesis, evolution and history. Four Courts Press, Dublin
McCarthy D, Breen A (1997a) An evaluation of astronomical observations in the Irish annals. Vistas Astron 41(1):117–138
McCarthy D, Breen A (1997b) Astronomical observations in the Irish annals and their motivation. Peritia 11:1–43
McCormick M (2011) History's changing climate: climate science, genomics, and the emerging consilient approach to interdisciplinary history. J Interdiscip Hist XLII:251–273
McCormick M, Büntgen U, Cane MA, Cook ER, Harper K, Huybers P, Litt T, Manning SW, Mayewski PA, More AFM, Nicolussi K, Tegel W (2012) Climate change during and after the Roman Empire: reconstructing the past from scientific and historical evidence. J Interdiscip Hist 43(2):169–220
McCormick M, Dutton PE, Mayewski PA (2007) Volcanoes and the climate forcing of Carolingian Europe, A.D. 750–950. Speculum 82:865–895
McGregor S, Timmermann A (2011) The effect of explosive tropical volcanism on ENSO. J Clim *24:2178*
Mitchell JK (2011) Looking backward to see forward: historical changes of public knowledge about climate hazards in Ireland. Ir Geogr 44(1):7–26
Mohtadi M, Prange M, Steinke S (2016) Palaeoclimatic insights into forcing and response of monsoon rainfall. Nature 533:191–199
Moriyama R, Sugiyama M, Kurosawa A, Masuda K, Tsuzuki K, Ishimoto Y (2017) The cost of stratospheric climate engineering revisited. Mitig Adapt Strat Glob Chang 22(8):1207–1228
Murphy D (ed) (1896) The annals of Clonmacnoise: being annals of Ireland from the earliest period to A.D. 1408: translated into English A.D. 1627 by Conell Mageoghagan. The University Press, for the Royal Society of Antiquaries of Ireland, Dublin
National Research Council (2006) Surface temperature reconstructions for the last 2,000 years. The National Academies Press, Washington, DC
Newfield TP, Labuhn I (2017) Realizing consilience in studies of pre-instrumental climate and pre-laboratory disease. J Interdiscip Hist 48(2):211–240
Nicholls K (1987) Gaelic society and economy in the high middle ages. In: Cosgrove A (ed) A new history of Ireland, II: Medieval Ireland 1169–1534. Clarendon Press, Oxford, pp 397–438
Noone S, Broderick C, Duffy C, Matthews T, Wilby RL, Murphy C (2017) A 250-year drought catalogue for the island of Ireland (1765–2015). Int J Climatol 37:239–254
Ó Corráin D (2005) Aspects of society, c.800. In: O Croinin D (ed) A new history of Ireland, vol 1, pp 549–608
Ó Cróinín D (1995) Early medieval Ireland. Longmans, London, pp 400–1200
Ogilvie AE, Farmer G (1997) Documenting the medieval climate. In: Hulme M, Barrow E (eds) Climates of the British isles: past, present and future. Routledge, London/New York, pp 112–134
Oppenheimer C (2011) Eruptions that shook the world. Cambridge University Press, Cambridge
Pausata FSR, Chafik L, Caballero R, Battisti DS (2015) Impacts of high-latitude volcanic eruptions on ENSO and AMOC. Proc Natl Acad Sci 112(45):13784–13788
Pfister C (2007) Climatic extremes, recurrent crises and witch hunts: strategies of European societies in coping with exogenous shocks in the late sixteenth and early seventeenth centuries. Mediev Hist J 10:33–73
Pfister C (2010) The vulnerability of past societies to climatic variation: a new focus for historical climatology in the twenty-first century. Clim Chang 100:25–31
Pfister C, Brázdil R, Barriendos M (2002) Reconstructing past climate and natural disasters in Europe using documentary evidence. PAGES News 10(3):6–7

Pilcher JR, Baillie MGL, Schmidt B, Becker B (1984) A 7,272-year tree-ring chronology for western Europe. Nature 312(5990):150–152
Robock A (2000) Volcanic eruptions and climate. Rev Geophys 38:191–219
Robock A (2008) 20 reasons why geoengineering may be a bad idea. Bull At Sci 64(2):14–18
Robock A (2014) Stratospheric aerosol geoengineering. In: Hester RE, Harrison RM (eds) Geoengineering of the climate system. Royal Society of Chemistry, Cambridge, pp 162–185
Robock A, MacMartin DG, Duren R, Christensen MW (2013) Studying geoengineering with natural and anthropogenic analogs. Clim Chang 121(3):445–458
Roy ED, Morzillo AT, Seijo F, Reddy SMW, Rhemtulla JM, Milder JC, Kuemmerle T, Martin SL (2013) The elusive pursuit of interdisciplinarity at the human–environment interface. Bioscience 63(9):745–753
Scheffran J, Brzoska M, Kominek J, Link PM, Schilling J (2012) Climate change and violent conflict. Science 336(6083):869–871
Schilling J, Freier KP, Hertig E, Scheffran J (2012) Climate change, vulnerability and adaptation in North Africa with focus on Morocco. Agric Ecosyst Environ 156:12–26
Schleussner CF, Donges JF, Donner RV, Schellnhuber HJ (2016) Armed-conflict risks enhanced by climate-related disasters in ethnically fractionalized countries. Proc Natl Acad Sci 113(33):9216–9221
Sigl M, Winstrup M, McConnell JR, Welten KC, Plunkett G, Ludlow F, Büntgen U, Caffee M, Chellman N, Dahl-Jensen D, Fischer H, Kipfstuhl S, Kostick C, Maselli OJ, Mekhaldi F, Mulvaney R, Muscheler R, Pasteris DR, Pilcher JR, Salzer M, Schüpbach S, Steffensen JP, Vinther BM, Woodruff TE (2015) Timing and climate forcing of volcanic eruptions for the past 2,500 years. Nature 523:543–549
Smyth WJ (2006) Map-making. In: Landscapes and memory: a geography of colonial and early modern Ireland c.1530–1750. Cork University Press, Cork
Stern PC, Ebi KL, Leichenko R, Olson RS, Steinbruner JD, Lempert R (2013) Managing risk with climate vulnerability science. Nat Clim Chang 3(7):607–609
Steinbruner JD, Stern PC, Husbands JL (eds) (2012) Climate and social stress: implications for security analysis. National Academies Press, Washington, D.C.
Stevenson S, Otto-Bliesner B, Fasullo J, Brady E (2016) "El Niño like" hydroclimate responses to last millennium volcanic eruptions. J Clim 29:2907–2921
Stoffel M, Khodri M, Corona C, Guillet S, Poulain V, Bekki S, Guiot J, Luckman BH, Oppenheimer C, Lebas N, Beniston M, Masson-Delmotte V (2015) Estimates of volcanic-induced cooling in the Northern Hemisphere over the past 1,500 years. Nat Geosci 8(10):784–788
Sweeney J (1997) Ireland. In: Mayes J, Wheeler D (eds) Regional climates of the British Isles. Routledge, London/New York, pp 254–275
Toohey M, Krüger K, Sigl M, Stordal F, Svensen H (2016) Climatic and societal impacts of a volcanic double event at the dawn of the middle ages. Clim Chang 136:401–412
Ulf Büntgen Ólafur Eggertsson Lukas Wacker Michael Sigl Fredrik Charpentier Ljungqvist Nicola Di Cosmo Gill Plunkett Paul J. Krusic Timothy P. Newfield Jan Esper Christine Lane Frederick Reinig Clive Oppenheimer (2017) Multi-proxy dating of Iceland's major pre-settlement Katla eruption to 822–823 CE. Geology 45:783–786
Zhang Z, Tian H, Cazelles B et al (2010) Periodic climate cooling enhanced natural disasters and wars in China during AD 10–1900. Proc R Soc 277B(1701):3745–3753
Zielinski GA (1995) Stratospheric loading and optical depth estimates of explosive volcanism over the last 2100 years derived from the Greenland ice sheet project 2 ice-core. J Geophys Res 100:D10.20.937–D10.20.955
Zielinski GA et al (1994) Record of volcanism since 7000 B.C. from the GISP2 Greenland ice-core and implications for the volcano-climate system. Science 264(5161):948–952
Zielinski GA et al (1996) A 110,000-Yr record of explosive volcanism from the GISP2 (Greenland) ice-core. Quat Res 45:109–118

Chapter 4
Film and the Medical Humanities: The 'Romantic Science' of Neurocinema

Germán Gil-Curiel and Armida de la Garza

Abstract Oliver Sacks and Alexander Luria advocated for 'a romantic science', a literary form at the intersection of fact and fable, which Sacks employed to introduce lay readers to the complexities of the brain, providing an excellent example of how the two cultures of science and humanities could be reconciled. This is the goal of the medical humanities, which emerged in recognition of the fact that medicine is an art just as much as it is a science. Here we argue there is a particular affinity between film and the brain that the medical humanities, which have hitherto mostly focused on literature, music and the fine arts, could fruitfully develop.

Keywords Neurocinema · Film and the brain · Medical humanities · Romantic science · Requiem

Oliver Sacks and Alexander Luria famously advocated for what they called a 'romantic science' (Sacks 1985 [2015], p. 184), a literary form operating at the intersection of fact and fable which Sacks employed in his vast oeuvre to introduce lay readers to the complexities of the brain, thus providing an excellent example of how the two cultures of science and humanities could be reconciled. This is ostensibly the goal of the medical humanities, which first emerged as an interdisciplinary field in the United Kingdom in the 1990s, in recognition of the fact that medicine is an art just as much as it is a science. In this context, we argue that there is a particular affinity between film and the brain that the medical humanities, which have hitherto mostly focused on literature, music and the fine arts,[1] could well fruitfully develop further. We first outline two ways in which film and medicine have been related within the medical humanities, namely, for pedagogical and theory

[1] Both visual arts and music are also used therapeutically, encouraging patients to engage in creative pursuits, particularly in psychiatry and in hospitals (Jordanova 2014, p. 43). It is not only consuming but also performing and producing the creative arts that relate medicine to the arts and humanities.

G. Gil-Curiel (✉) · A. de la Garza
College of Arts, Celtic Studies and Social Sciences, University College Cork, Cork, Ireland
e-mail: german.gc@mac.com

© Springer International Publishing AG, part of Springer Nature 2019
A. de la Garza, C. Travis (eds.), *The STEAM Revolution*,
https://doi.org/10.1007/978-3-319-89818-6_4

development purposes, and outline a new paradigm, the biocultural, more concerned with physiological intervention. We then discuss memory, schizophrenia and hallucination in connection to film and illustrate these with a discussion of the film 'Requiem' by Alain Tanner (1998). We finish the chapter with concluding remarks.

Film and the Medical Humanities Project

Initially, the aim of the medical humanities was primarily the pursuit of educational goals, to restore to students 'many common-sense qualities of empathetic understanding that they come to lose in the course of medical training [...] and which can be summarised as empathy, reflection and trust' (Greco 2013, p. 234), enacting in this way the logic of accountability that is often ascribed to interdisciplinarity (i.e. interdisciplinarity for social relevance and engagement). Narrative was crucial since fiction allows medical students 'to enter the worlds of their patients, if only imaginatively, and to see and interpret these worlds from the patients' point of view' (Whitehead 2014, p. 107). In the case of cinema, the evolution of ethical attitudes to medical dilemmas such as mercy killing and abortion can be seen on film, as well as depictions of illnesses and of medical practitioners, underscoring the changing regard in which societies have held their professions and eliciting discussion and reflexion. Coma, stroke, meningitis, encephalitis, multiple sclerosis, seizure and dementia are among the neurologic disorders that have been depicted on film. And the contention that 'the secret of being a good doctor is act[ing] like one' (Wijdicks 2015, p. 3) puts forward the performative dimension of the practice. Many have found in films elementary scripts to cue in action.

In addition, the medical humanities have sought to highlight the similarities between the medical epistemologies and the interpretive, hermeneutic ways of knowing that are traditional of the humanities, arguing that a practice 'is often at its most rational when it does not exclude subjectivity' (Greco 2013, p. 236). Theories traditionally employed in the humanities, such as postcolonial studies, have been innovatively deployed to account for the relations between patients and medical practitioners in a postmodern context characterised, as put by Monica Greco, by chronic rather than acute illness, in which patients lack a voice: 'Just as colonised peoples claim recognition of what their labour has contributed to the prosperity and civilisation of colonising powers, the ill are demanding [...] that medicine recognise its need for them' (p. 235), not least through case history in clinical texts, and the calls for volunteers in randomised control trials, among others. By the same token, developments in film theory during the 1970s were informed by psychology and psychoanalysis, and later by cognitivism, and conversely the study of perception that is today the concern of neurology has been informed by the technology of film. It has been widely acknowledged that the models and metaphors employed by scientific theories to represent the phenomenon under study condition the scientific imagination and have important epistemological dimensions since they 'enable us to see aspects of reality that the metaphor's production helps to constitute', narrowing the distance between science and the humanities since from this perspective 'both are affairs of the imagination' (Guldin 2016, pp. 12–14). The zoetrope[2] provided one of the first metaphors of the brain in order to theorise consciousness, considered discontinuous, a succession of separate sensations and images like those of the zoetrope that only resembled a flow on account of an illusion that connected them together, provided by the brain. Moreover, the technical and conceptual devices of cinema such as zooming, fading, dissolving, allusion and juxtaposition have been acknowledged to closely resemble the 'stream' of consciousness (Sacks 2017). Certain forms of brain malfunction such as seizures or migraine or the use of drugs lend credence to the view that consciousness is indeed discontinuous, since perception in such cases is experienced as a series of discrete moments. And because focusing on any given perception out of the thousands available is an active and selective process that depends on our memories and feelings—and thus highly subjective—an analogy between the perceiving subject and a film director seems pertinent: 'we are the directors of the film we are making, but we are its subjects too: every frame, every moment, is us, ours' (Sacks 2017).

Apart from the pedagogical and theory development applications of the medical humanities, a biocultural approach to cinema has recently emerged in which methods and tools from the laboratory are combined with materials and artworks from the arts. Functional magnetic resonance imaging (fMRI) and PET scans focusing on cortical activity have started mapping the network of brain parts that are activated when viewing narrative sequences on film (Young 2012, p. 108). Along with computerised neural modelling using networks of virtual neurons that organise themselves in response to a variety of stimuli and constraints, these techniques are today enabling the study of the physiological effects that watching films and nonnarrative moving images have on the brain, opening the possibility of their eventual use as treatment.[3] For as put by Torben Grodal, 'When we watch a film, our heart rhythms change, we sweat, and our muscles alternately tense and relax [...] these bodily changes are linked to emotional reactions that also play a central role in memory, cognition and consciousness' (Grodal 2009, p. 4), on account of mirror neurons, which allow viewers to experience themselves an action viewed (Smith 2014, p. 34). In experiencing an emotion, a plethora of interconnected neuropsychological changes 'sweep through the brain and the body [...] hormonal and neuro-chemical processes are integral in shaping affect' (Robertson 2014, p. 239). Indeed, because emotions are now not conceptualised as binary (i.e. as existing or not in a given time) but rather 'as a process which is dynamically shaped vis-à-vis environmental and internal factors' (Raz et al. 2013, p. 286), films, which unfold over time, provide the ideal means to empirically research the neural correlates of emotion and have

[2]A pre-film animation device that produces the illusion of motion by displaying a sequence of drawings or photographs showing motion in its progressive phases

[3]At the moment, however, these findings are being capitalised upon by the Hollywood film industry, which increasingly relies on fMRI techniques as opposed to the previously widely used focus groups, to monitor audience's reactions to specific scenes in films and indeed to whole films, for creative and marketing purposes (Randall 2011).

become the favourite method employed by researchers because they excel in capturing attention and triggering emotion.

Video has also been successfully employed for the support of (left) hemi-field extinction resulting from stroke, in which case patients lose all awareness of the existence of the left side of their bodies. With a camera and monitor both facing the patient, it is possible to turn the screen into a mirror of sorts, so that the left side appears on the right, showing much promise as a form of video feedback for patients (Sacks 1985 [2015], p. 82).

Film and Memory: Schizophrenia, Hallucination and Dream

Being able to remember has profound philosophical implications, since our sense of self, identity and personhood is closely entwined with our memory (Foster 2009, p. 94). Memory requires the brain to be physically altered by experience. While a full account of the process is beyond the scope of this chapter, suffice it to say here that this requires the formation of new synapses (i.e. connections between neurons), facilitated by puffs of serotonin, a neurotransmitter found in all animals which strengthens a synaptic connection in both the short and the long term, depending on the amount 'puffed' upon the connection (O'Shea 2005, p. 95). Synaptic change, or plasticity, is thus fundamental to memory formation and learning. The metaphor of a library is often employed to describe memory, with the hippocampus, the part of the brain involved in its production, likened to a 'printing press' for new memories, which are then 'filed' as 'books' for access (retrieval) in the cerebral cortex (Foster 2009, p. 84). On the other hand, the formation of new memories leads to an ongoing recategorisation, updating and re-correlating, in such a way that remembering is a constant process of recreation. Emotional events also have an impact upon memory, and situations of threat or reward tend to enhance the process—the so-called 'flashbulb' memory (Foster 2009, p. 61). Whether we experience events in real life or watch them on film, we build perceptual and memory representations in the same format. It has thus been argued that when information from a film is plausible and similar enough to information from real experience and if enough time has gone by, they might indeed be confused. For this reason, the ability to build event representations by means of film viewing has been regarded as a form of behaviour that fosters adaptation: by watching films we can construct memories 'on the cheap, that we can use to get around in the world' (Zacks 2014, p. 111).

There are instances of people who enter dissociative states, such as in schizophrenia, in which they become partly or completely separated from their memories. Studying how exactly this happens has been important to be able to discriminate between different types of memory processes and to link deficits in remembering with specific neurological structures, for 'with complex systems such as the brain we learn more about the functional relationships […] when they cease to function properly than when everything is working smoothly' (Foster 2009, p. 89). Generally speaking, schizophrenia implies a breakdown in the signifying chain: if we are unable to unify the past, present and the future of the sentence, then we are 'similarly unable to unify the past, present, and the future of our own biographical experience of psychic life. With the breakdown of the signifying chain, therefore, the schizophrenic is reduced to an experience of pure material signifiers, or, in other words, a series of pure and unrelated presents of time' (Jameson 1991, p. 27), the frozen frames of the zoetrope.

Fredric Jameson famously extrapolated these insights to theorise artistic production in postmodernity,[4] a period characterised by profound changes in our collective experience of time, partly due to technological change.[5] In short, he contended the schizophrenic feature at the heart of the postmodern condition is evident in an acute and emotional state of the mind that he described as 'a whole new type of emotional ground tone [...or] "intensity". A disturbing sense of unreality in the shape of an extreme experience in which there is a sudden collapse of temporality, which releases the present from all activities and intentionalities that might focus it and makes it 'a space of praxis' (Jameson 1991, p. 27). Thereby isolated, that present suddenly engulfs the subject in an extremely vivid fashion, even overwhelming in the materiality of its perception.

Compulsive access to specific memories can also occur, as if the normal process of revision and recategorisation mentioned above failed to apply to those and they became 'fossilised' or 'petrified' as a result, in what has been called the interictal personality syndrome. By contrast, cases have also been recorded of patients whose illnesses lead their brain to gain seemingly unrestricted and random access to memories, some of which they had long forgotten. A patient suffering from astrocytoma with frequent temporal lobe seizures characterised by dreamy states and involuntary reminiscence without loss of consciousness described images of family, neighbours and landscapes from her home village, seemingly in the state of dream madness that is called oneirophrenia, which sometimes occurs in schizophrenia (Sacks 1985 [2015], pp. 162–163). However, these 'phantasms'[6] were all memories, 'doubling' her consciousness, and 'they seemed more like certain paintings or tone poems'.

---

[4]According to Jameson, in society a sort of schizophrenia has eroded historicity 'in the new forms of our private temporality' and has established 'new types of syntax or syntagmatic relationships in the more temporal arts' (Jameson 1991, p. 6). However, Jameson uses Lacan's account of schizophrenia as a description rather than a diagnosis, insofar as 'it seems [...] to offer a suggestive aesthetic model' (p. 26).

[5]Ever since personal computers became widespread at the turn of the century, an increase in narrative complexity in popular cinema has taken place, a trend variously termed 'modular narrative' or 'database aesthetic' (Cameron 2008, p. 1). These films articulate a sense of time as divisible and subject to manipulation, essentially as consisting of discrete, segments that can be accessed in a non-linear manner, in the same way that files stored in a computer—or memories in a brain. See, for instance, 'Eternal Sunshine of the Spotless Mind' (2004), 'Vanilla Sky' (2001) or 'Memento' (2000).

[6]Many cultures regard hallucinations and dreams as privileged states of consciousness, actively sought through diverse means such as drugs, meditation or isolation. Western cultures however ascribe negative connotations to hallucinations, including madness or brain damage or malfunction, and admitting to them carries a stigma. But hallucinations may well be at the origin of art. The geometric patterns of migraine might prefigure the motifs of aboriginal art and also the visions of

Requiem

This is exactly the case of the main protagonist of the film 'Requiem' by Alain Tanner (1998), after the novel by Antonio Tabucchi entitled 'Requiem: a Hallucination' (1994), set in Lisbon. Although described as a 'heavy on atmosphere and 100% *cerebral* picture [that …] offers little to rouse the enthusiasm of even art house regulars unless they are planning a trip to Lisbon' (Young 1998, emphasis ours), we claim that from the perspective of neurocinema, 'Requiem' provides an excellent figurative entrance into the mind of a protagonist that may be suffering from schizophrenia and hallucination or from an experience such as the one that the patient with astrocytoma described above had. But while the clinical disciplines are 'deeply invested in delineating the aberrant, diseased, dysfunctional or distressing from what we might call "normal"' (Woods and Fernyhough 2014, p. 84) and most genre cinema subscribes to this representation—especially in science fiction—the lack of fixed conventions in art cinema and the subjectivity that the shooting style conveys value all experience and take it seriously, including the hallucinatory experience, depicting it as meaningful. Normalisation or, in other words, recognising and valuing the diversity of experience as part of what makes us human is arguably an important contribution of the arts, in this case art cinema, to the realm of medicine.

The narrative centres on Paul, a writer who has a rendezvous with a friend and fellow writer, the Portuguese poet Fernando Pessoa, who is no longer among the living and who can quite plausibly be interpreted as Paul's alter ego since one of the fundamental archetypes of schizophrenia is the collapse of identity, closely associated, in art, with the everlasting tradition of the doppelganger (in the novel too, the character of Paul is sometimes interpreted as author Antonio Tabucchi's alter ego). It is a circular story, which starts at midday and concludes at midnight. It is also a hallucinating oneiric experience in which barriers of conventional time fall apart to allow the people of the present and past to meet and communicate. As an allegory of the vicissitudes of the self, in 'Requiem' the return of ghosts becomes the nucleus of the narrative.

Requiem is a journey ridden with ambiguity and uncertainty. Paul is in search of himself via reliving fragments from his past, and he passes the day 'meeting long-dead friends from his memories, who materialise out of nowhere with complete naturalness' (Young 1998). This is quite similar to the way in which the theatrical film-viewing experience was initially conceptualised, that is, as meeting 'shadows', in a way long dead, that materialised seemingly out of nowhere, which viewers would engage on their own personal terms depending on previous experience. Film director and surrealist artist Luis Buñuel indeed famously characterised film as

light or halos that viewers attributed to saints and apparitions. Lilliputian hallucinations may account for imps, elves and fairies, and 'ecstatic' seizures may play a role generating a sense of the divine (Sacks 2012, p. xii).

inherently germane to dream.[7] Among the people Paul encounters is his friend Pierre, a poet with whom he had a romance in common with a woman called Isabel; Isabel herself, who committed suicide, apparently out of remorse following the abortion of a child that may have been Pierre's or perhaps Paul's; and Paul's own father, who pays him a visit while in a dream in which he appears as a young sailor and demands that Paul recount to him the way he died. Finally, his journey culminates in the meeting with his friend, the ghost of the late Pessoa.

Illness, suffering and pain, including the pain arising from the sense of loss that death brings, constitute a large part of the recollections involved, not surprisingly since remembering is to a large extent mediated by the emotional commitment to and investment in the event (Foster 2009, p. 12), and also given that the need to bridge or reconcile separations from people or places, or discontinuities from events or situations is the cause of both unusual brain activity and the need to symbolically repair that is at the root of much artistic work. The protagonist's encounter with his father is related to death and agony because of the extreme physical and mental suffering his father endured. When he asks his son how his life will end, Paul replies he will die/died as a result of a long and painful disease, partly caused by medical negligence during an operation as described on the film.

Other artworks the film presents are music and painting. A requiem is both a mass offered for the dead and the piece of music that is played at these ceremonies. Music is of course a powerful means to evoke memories. In 'Requiem', Paul's memories of his stay at a now dilapidated lighthouse are awoken by music. He calls to mind those times in front of the rickety piano when he 'sat down and with one finger, from memory, […] played the melody from a nocturne by Chopin'. As Paul now tries to play the piece, he finds the piano is broken, but his procedural memory, i.e. the embodied memory that comes from physical performance and survives even in cases of amnesia, is intact. As for painting, close to the end, Paul visits the *Museum of Ancient Art* to see *The Triptych of Temptation of St Anthony* by Hieronymus Bosch, dating from around 1501. The painting tells the story of the spiritual torments endured by Saint Anthony Abbott throughout his life. This visit gives rise to a reflection on the nature of art. First, there is reflection on its capacity for reconstructing and restoring its meaning throughout time, by means of the viewer's perception. Pondering on whether the painting has remained the same, the protagonist maintains it actually had changed because his own perspective had already changed, through a process not unlike the continuous updating of memories in the brain, which reclassifies them while ascribing to them to new meanings. Moreover, the extravagant fantastic of the painting gives rise to two contrasting points of view about the artist: on the one hand, a painter who is copying the triptych argues that

[7] As the character of Pessoa puts it later in the film 'if "the evocation" has the power to recall the dead, if its faculties of medium allow them to recall the deceased, it is because it is also a *convocation*. The image of the deceased appears and materialises thanks to the [director], it returns to life: we are in the presence of a ghost. The voice [of the director] has the power to establish a dialogue with the ghosts' or, from our perspective, with memories (Tabucchi, *Requiem*, 165–166, my translation).

the artist, having a 'perverse imagination', created a personal interpretation that exceeds by far the saint's torments; in contrast, the protagonist thinks that Bosch would have painted 'the storm in the saint's delirium'. The relation between art and medicine is also alluded to by the copyist apropos Bosch's triptych, which had formerly possessed the virtue of curing disease, as 'sick people would file past it hoping that some miraculous intervention would put an end to their suffering'. Indeed, the painting has been exhibited at the St. Anthony Hospital of Lisbon.

According to Jameson, extreme instant of schizophrenia can best be grasped by a return to older theories of the sublime (p. 6). The sublime was associated by Edmund Burke to the limits of terror and an extreme state of mind and emotion so powerful that it would make human beings collapse in an instant. Parallel to the main theme of death and nostalgia for the past times, 'Requiem' is imbued with a sense of the sublime by means of the strong yet subtle emotions that the encounters convey. This overwhelming feeling reaches its climatic point when the protagonist meets Isabel, a crucial presence in his past. This scene delivers emotional and poetic reconciliation between the three persons involved, Paul and the two shadowy presences of Isabel and Pierre, through a waltz that they dance together. The denouement—conceived as sublime—shows that the collapse of identity leads to reconstitution, reconciliation and death of the self, respectively. Paul, unable to unify the temporality of the biographical experience or his psychic life, can live only a collapsed temporality that is all present and makes of the people that he meets in Lisbon the 'space of praxis' that Jameson alludes to.

Conclusion

This chapter has briefly sketched out the spaces where the study of medicine and the arts and humanities intersects, with a focus on film and brain science. These spaces have mainly been the pedagogical, with the use of films as teaching aids, and, at the deeper level of epistemology, the innovative application of theories from one field into the other. We have then outlined a third—and in our view more fruitful—paradigm, namely, the collaboration between film studies and brain science for physiological intervention, a field that shows much promise in advancing our understanding of emotion, memory and cognition in general. We ended with a demonstration of how a medical humanities framework can be useful in the analysis of 'Requiem', an art film understood in this light as portraying schizophrenia, hallucination and a compulsive access to memory from a deeply humanising perspective, as opposed to the so-called medicalisation approach characteristic of the health sciences and which much science fiction cinema reifies.

We note that from an interdisciplinary perspective, the first two approaches in particular, namely, the pedagogical and the theory-building uses, have been regarded as merely additive at best and in any case as a stage to be superseded by an integrative approach in which 'the nature, goals and knowledge base of clinical medicine itself might be challenged and reshaped by its encounters with the humanities'

(Whitehead 2014, p. 108). We however find this characterisation of the pedagogical and theory-building uses of the medical humanities misleading, for these and other more negative terms used to describe interdisciplinary research in general such as 'trespassing', 'poaching' or 'indulging in creative parasitism' or even 'promiscuity' (Osborne 2013, pp. 86–91) are discipline-centric and lack a historical perspective of how disciplines originated in the first place, taking them for granted.[8]

Drawing from Michel Foucault's investigation of the process of disciplinary formation, Roberto Pacheco et al. argue that just as the scientific disciplines evolved in the nineteenth century to obtain strict subjection of a body of knowledge (2017, pp. 307–308), in a time in which analogue media allowed limited access to the production and dissemination of knowledge, the present time in which digital media have dramatically expanded these, can instead be characterised as a moment of experimentation with new combinations and structures, a dynamic of innovation through creative destruction. Whether described positively as 'collaboration' or negatively as with the terms mentioned by Osborne above is immaterial. The outcome will most likely be problem or issue-centric and post-disciplinary.

Be that as it may, what is relevant to us is that, as Marina Roseman puts it, up to now allowing our respective disciplines 'to remain opaque to one another has historically served a segmented political design of institutionalised knowledge and power' (Roseman 2011, p. 20), which is totally unnecessary. Indeed, she continues, 'given the weight of suffering around us, on the one hand, and the amount of knowledge available to us on the other, it may not be morally viable' to continue to do so. Far beyond what they can offer each other, various combinations of film studies and neuroscience have much to offer to patients, their families and society in general.

Works Cited

Callard F, Fitzgerald D (2014) Medical/Humanities. In: Holden J, Kieffer J, Newbigin J, Wright S (eds) Where does it hurt? The new world of the medical Humanities. Wellcome Trust, London, pp 16–17

Cameron A (2008) Modular narratives in contemporary cinema. Palgrave, London

Crowe C (Director) (2001) Vanilla Sky [Motion picture]. USA

Foster JK (2009) Memory: a very short introduction. Oxford University Press, Oxford

Gondry M (Director) (2004) Eternal sunshine of the spotless mind [Motion picture]. USA

Greco M (2013) The logics of interdisciplinarity: The case of medical humanities. In: Barry A, Born G (eds) Interdisciplinarity: reconfiguration of the social and natural sciences. Routledge, London/New York, pp 226–246

Grodal T (2009) Evolution, emotion, culture and film. Oxford University Press, Oxford

Guldin R (2016) Translation as metaphor. Routledge, London/New York

Jameson F (1991) Postmodernism or the cultural logic of capitalism. Verso, London/New York

[8] Felicity Callard and Des Fitzgerald offer a more nuanced approach: they do not see the medical humanities as a meeting point for demarcated territories, but see them instead as a series of knowledges, materials and practices mixed a priori 'whose ongoing embroilment is entirely indifferent to covetous claims regarding disciplinary contribution' (2014, p. 16).

Jordanova L (2014) Medicine and the visual arts. In: Bates V, Bleakley A, Goodman S (eds) Medicine, health and the arts: approaches to the medical humanities. Routledge, London/New York, pp 41–63

Nolan C (Director) (2000) Memento [Motion picture]. USA

Osborne T (2013) Inter that discipline! In: Barry A, Born G (eds) Interdisciplinarity: reconfigurations of the social and natural sciences. Routledge, London/New York, pp 82–98

O'Shea M (2005) The brain: a very short introduction. Oxford University Press, Oxford

Pacheco RC, Manhaes M, Uriona Maldonado M (2017) Innovation, interdisciplinarity and creative destruction. In: Frodeman R (ed) The Oxford handbook of interdisciplinarity, 2nd edn. Oxford University Press, Oxford, pp 303–318

Randall K (2011, February 25) The rise of neurocinema: how hollywood studios harness your brain waves to win Oscars. (Fast company) Retrieved May 18, 2015, from http://www.fastcompany.com/1731055/rise-neurocinema-how-hollywood-studios-harness-your-brainwaves-win-oscars

Raz G, Hagin B, Hendler T (2013) E-motion pictures of the brain: recursive paths between affective neuroscience and film studies. In: Shimamura AP (ed) Psychocinematics: exploring cognition at the movies. Oxford University Press, Oxford, pp 285–313

Robertson P (2014) Music, therapy and technology. In: Bates V, Bleakley A, Goodman S (eds) Medicine, health and the arts. Routledge, London/New York, pp 227–245

Roseman M (2011) A four-fold framework for cross-cultural, integrative research on music and medicine. In: Koen B (ed) The Oxford handbook on medical ethnomusicology. Oxford University Press, Oxford, pp 18–45

Sacks O (1985/2015) The man who mistook his wife for a hat. Croydon: Picador

Sacks O (2012) Hallucinations. Picador, London

Sacks O (2017) The river of consciousness. Picador, London

Smith M (2014) The pit of naturalism: neuroscience and the naturalized aesthetics of film. In: Cognitive media theory. Routledge, New York/London, pp 27–45

Tabucchi A (1994) Requiem: a hallucination (trans: Costa MJ). New Directions Publishing Corporation, New York

Tanner A (Director) (1998) Requiem [Motion picture]. Switzerland, France, Portugal

Whitehead A (2014) The medical humanities: a literary perspective. In: Bates V, Bleakley A, Goodman S (eds) Medicine, health and the arts: approaches to the medical humanities. Routledge, London/New York, pp 107–127

Wijdicks EF (2015) Neurocinema: when film meets neurology. CRC Press, London/New York

Woods A, Fernyhough C (2014) Hearing voices. In: Holden J, Kieffer J, Newbigin J, Wright S (eds) Where does it hurt? In my soul. Wellcome Trust, London, pp 84–85

Young D (1998, May 01) Requiem. Retrieved November 1, 2017, from variety: variety.com/1998/film/reviews/requiem-2-1200454193/

Young S (2012) Psychology at the movies. Wiley-Blackwell, Oxford

Zacks J (2014) Flicker: your brain on movies. Oxford University Press, Oxford

Chapter 5
Heritage Science: A Report

Brendan Dooley and Armida de la Garza

Abstract Heritage Science is a growing interdisciplinary field that illustrates the benefits of STEAM in action. It attempts to join specific competences in material science with the interpretative skills of the humanities and social sciences in order to understand and assist with the conservation of artefacts and sites that different cultures and societies value, while also assisting with the curation of tangible artistic and archaeological capital of interest from economic and development perspectives. This chapter presents a report of a 2-day symposium that took stock of the interdisciplinary field, presented existing projects and introduced new avenues for future research.

Keywords Heritage science · Heritage conservation · Heritage · Archaeological capital · Interdisciplinary collaboration

Heritage science is a growing interdisciplinary field that very aptly illustrates the benefits of STEAM in action. In brief, it attempts to join specific competences in material science with the interpretative skills of the humanities and social sciences in order to understand and assist with the conservation of artefacts and sites that different cultures and societies value while also assisting with the curation of tangible artistic and archaeological capital of interest from economic and development perspectives. As national heritage sites and artefacts often seem to be on a race between preservation and destruction, increasing awareness of the threats posed by natural and anthropogenic forms of damage, along with a perception of the tangible features of cultural distinctiveness, has added urgency to protecting and recording that which may one day disappear forever. More and more scholars in humanities and natural science fields in Europe see this diverse and interdisciplinary field as an ideal means to cater for their needs. Along with the European Research Infrastructure for Heritage Science led by Italy, other structures are emerging, such as access,

B. Dooley
University College Cork, Cork, Ireland

A. de la Garza (✉)
College of Arts, Celtic Studies and Social Sciences, University College Cork, Cork, Ireland
e-mail: adelagarza@ucc.ie

© Springer International Publishing AG, part of Springer Nature 2019

A. de la Garza, C. Travis (eds.), *The STEAM Revolution*,
https://doi.org/10.1007/978-3-319-89818-6_5

research and technology for the conservation of the European Cultural Heritage (EU-ARTECH), Cultural Heritage Advanced Research Infrastructures: Synergy for a Multidisciplinary Approach to Conservation/Restoration (CHARISMA), Integrated Platform for the European Research Infrastructure ON Cultural Heritage (IPERION CH) and Advanced Research Infrastructure for Archaeological Dataset Networking in Europe (ARIADNE). The new journal entitled *Heritage Science* provides specific examples of the exciting promise that this arts and science collaboration offers:

> By examining the pigments in a painting, can we tell how it was painted, where it was painted and even whether it is a forgery? By looking at dyes in a textile can we find out about the origin of manufacture and the geographical route it took to get to its destination and hence about trade routes? By analysing the metal content of a coin, can we tell about the economic factors of the time and even whether it was created from melted down coins from a neighbouring country? (Brereton 2013)

To answer such questions, scientific approaches provide additional insight. One might add paleopathology, the analysis of historic remains to test for disease and other causes of death. The use of modern computer techniques, together with the knowledge of forensic science that is these days often displayed on television programmes, has started to familiarise the general public with heritage science. See, for instance, the account of maritime archaeologist Margaret Rule who explains how soil analysis and remote prospection using ground-penetrating radar or a magnetometer enabled researchers to present an accurate reconstruction of a lost eighteenth-century garden, which was then shown on television (Rule 2006, p. 1).

There is no denying that there has been a slow but steady movement towards the adoption of scientific methods in cultural studies, not to mention in literary history, historiography and archaeology. Moreover, 'new instrumentation and approaches are required to study objects in museums, posing special challenges, for example miniaturisation to bring non-invasive measurements into the museum shelves, screening large numbers of objects, finding out about hidden layers in paintings, checking for possible decay' (Brereton 2013). All of these require methods that are tailored to the specific needs of cultural conservation.

A 2-day symposium and masterclass was held at University College Cork (UCC) on 2–3 November 2017 to explore the challenges and opportunities of heritage science, with a view to demonstrating the ways in which specific projects and interests may be enhanced and developed when viewed within a heritage science perspective.

The Symposium

Presentations started with Professor Brendan Dooley from Digital Arts and Humanities at UCC, who introduced the project 'The Venice Time Machine'. This project aims to build a virtual reality representation of the city of Venice that will be fully cross-referenced to a digital archive of the city's cultural heritage that covers more than a thousand years of its history.[1] When completed, it will show the way that news, money and commercial goods circulated in the city and point to migration and artistic/architectural patterns while also becoming an example of the way in which such big data applications may be used for other cities in the future.

The keynote speech was provided by Professor Matija Strlič from University College London. In addition to providing a broad overview of the state of the field, Professor Strlič presented recent research into the smell of old paper and the use of volatiles as markers for degradation. The research involves a study of the various chemical substances that cause the characteristic smell of old books and recreating this smell in the laboratory for its application in museums and libraries special collections in order to add a sensory dimension to visitors' experience. The research, co-authored by Cecilia Bembibre, has attracted considerable media interest and received extensive coverage, including in the form of a video.[2]

Another highlight to the symposium was provided by Daniela Iacopino, an expert in nanomaterials from Tyndall Research Institute with a great interest in inks. Following a fascinating introduction on the nanoparticles of gold and its various colourings, which can include red hues, she then discussed the many ways in which applications of nanoparticles are helping with the restoration and preservation of paintings and drawings. Among the examples Iacopino provided was the case of a painting that had been defiled by graffiti, which was restored through the application of a nano-gel that had been chemically modified to selectively dissolve the graffiti layer, leaving the painting underneath intact. A gel employing a different formulation was similarly used to clean traces from cellotape on a number of modern drawings. And nano-cellulose applied to paper, either made from wood pulp or linen, can help to restore tears and damage caused by the passage of time.

Two presentations in the symposium were devoted to issues related to the conservation of heritage books. The first was delivered by Crónán Ó Doibhlin, Head of Research Collections and Communications at Boole Library in University College Cork, Ireland, who discussed conservation and preservation issues in the context of library resources. A resource like the Great Book of Ireland,[3] a manuscript anthology of modern Irish art and literature acquired in 2013 and held in Special Collections at Boole Library, presents special problems of conservation, although it was only made in 1989–1991. But even in such a short period of time, some of the work included there has deteriorated and in certain cases become illegible, challenging the current goal of keeping the book for a thousand years as a testimony of Irish heritage.

[1] See 'A Virtual Time Machine for Venice' at https://www.youtube.com/watch?v=uQQGgYPRWfs

[2] The video is entitled 'Smell of Heritage: The Historic Book Odour Wheel' and can be viewed here: https://www.youtube.com/watch?time_continue=123&v=7gO4jaTmAz0. The paper is Cecilia Bembibre and Matija Strlič, 'Smell of heritage: a framework for the identification, analysis and archival of historic odours', *Heritage Science* 5:2 (2017).

[3] For more on the Great Book of Ireland, see: https://www.youtube.com/watch?v=EuOyE0i16pE

Professor Pádraig Ó Macháin, Head of Modern Irish at UCC, then explored the thousand-year-old tradition of Irish manuscript production, uninterrupted until the development of reliable Irish printing in the nineteenth century. Due to the vicissitudes of Irish collections around the world, digitization has been a particular boon, such that technology now allows comparison between exemplars and the joining of separated fragments in ways hitherto impeded by distance between repositories when not prevented by the sheer fragility of the material.

The last presentation by Mary Teehan from the Discovery Programme in Dublin involved an introduction to current and past projects in Ireland, such as WODAN (Wood and Charcoal Database); LIARI (Late Iron Age and 'Roman Ireland'); 3D ICONSl, described as 'a collection of highly accurate 3D models, images, texts and videos of over 130 iconic and internationally important monuments and buildings from Ireland'; and CHERISH (Climate, Heritage & Environments of Reefs, Islands and Headlands). She pointed out her role in the new ERIHS network (European Research Infrastructure for Heritage Science) while directing attention to the challenges of inter- and transdisciplinarity, in fields and studies where stakes are so high not only for the natural, biological and human sciences and humanities but also for research institutions and governments.

The symposium ended with a discussion of the various meanings of the word 'heritage', in the light of the specific collaboration between art and science that 'heritage science' is arising. In the future, other symposia will aim to cover fields that it was not possible to include this time, such as the conservation of film either in celluloid or in digital form, and also the use of genetic modification of plants to preserve and develop the natural heritage of botanical gardens and national parks.

All in all, the symposium demonstrated many concrete results of current and past interdisciplinary collaborations between and among arts and science.

References

Brereton RG (2013) Editorial. Herit Sci 1(1):1

Rule M (2006) Introduction. In: May E, Jones M (eds) Conservation science: heritage materials. The Royal Society of Chemistry, Cambridge, pp 1–12

Part II
Science and Art: Reflective Essays by Scientists and Artists

Chapter 6
In Between Science and The Arts: Dancing a PhD in Renewable Energy

Aoife Long and Armida de la Garza

Abstract The arts are an arena in which the body is central to the process of inquiry and indeed constitutes a mode of knowing. In a traditional academic setting, engineering and the arts are considered separate fields. This dance project aims to bridge the gap between these fields and between academia and wider society through a participatory approach to expressing engineering knowledge. Participants were invited to convey the knowledge, and viewers were entertained by academic information presented in an alternative medium. This chapter outlines the approach taken to produce the film, from initial conception of the idea to planning and filming.

Keywords Embodiment · Dance · Science communication · Renewable gas

Introduction

I first learned about the 'Dance your PhD' competition as a result of a social media training programme for researchers that was held at the Marine and Renewable Energy Research Institute (MaREI) at University College Cork in Ireland (UCC), where I am currently a PhD candidate.[1] The programme focused on how to build a social media profile, the work required to effectively communicate science research using social media, and explained why it could be valuable for researchers. I set up

---

[1] Open to PhD candidates in biology, physics, chemistry and social science, the 'Dance your PhD' competition was first launched in 2007 and is held annually, sponsored by Science Magazine, the American Association for the Advancement of Science and HighWire Press. A panel consisting of both artists and sciences selects the winners. In 2017 it offered a cash prize of $2500 USD, divided among the four categories. Winning videos can be viewed at http://www.sciencemag.org/news/2017/05/announcing-2017-dance-your-phd-contest

A. Long (✉)
MaREI Centre, Environmental Research Institute, University College Cork, Cork, Ireland

School of Engineering, University College Cork, Cork, Ireland
e-mail: aoife.long@ucc.ie

A. de la Garza
College of Arts, Celtic Studies and Social Sciences, University College Cork, Cork, Ireland

© Springer International Publishing AG, part of Springer Nature 2019

A. de la Garza, C. Travis (eds.), *The STEAM Revolution*,
https://doi.org/10.1007/978-3-319-89818-6_6

my twitter account with a professional profile and found out about the competition through this platform, from a tweet by the European Union Science Communication hub, which promotes science communication in Europe under @EUSciComm. The tweet announced the 10th annual Science Magazine 'Dance your PhD' competition. Its aim was 'to encourage scientists to explain their research topic to a general audience with a video that should delight and inform the public' (Science 2017). Participation was to be judged by artistic merit, scientific merit and the creative combination of both. Text could be used to narrate, but the excessive use of text was to be penalised, so it was necessary to be as concise and imaginative as possible with the description of each scene. It was immediately apparent to me that this was a brilliant way to disseminate scientific research, and I started thinking of an outline for my research straight away (see Image 6.1).

My professional and academic background had previously alerted me to the value of collaboration between science and the arts. I graduated from UCC with a degree in Civil and Environmental Engineering in 2004. Since then I have worked for nearly 10 years as a structural engineer, a traffic engineer and a transport planner. In 2013 I briefly left engineering and worked as an Information Management Officer for the Child Protection and Education sectors of UNICEF in Lebanon, for 3 years. At the strategic level that UNICEF works, child protection is very much focused on behavioural change, and most of the practitioners had a social science or psychology background. It then became evident to me that these members of staff had the tools, understanding and language that are required to elicit behavioural change, which engineers in transport planning lacked. Most of the guidance for engineers on behaviour change is based on observation and studies by engineers. To adequately engage with the needs of child protection, cross-disciplinary training was necessary. Moreover, behaviour change is also very important for other engineering applications in humanitarian response and development. For example, engineers may design and build a water point, but if a community does not value or trust the source of clean water, then it simply will not be used. For this reason, the name of the sector responsible was changed from WatSan, standing for Water and Sanitation, to the acronym WASH, meaning Water, Sanitation and Hygiene Promotion. Through this kind of experience and observation, I became convinced that engineers need to work with and learn from other disciplines. The 'Dance your PhD' competition then posed an excellent opportunity for me to attempt to do this at an early stage of my research career as I am a first year candidate.

Moreover, as a female engineer, I am aware that I am part of a minority, and I believe that the profession would benefit from being more diverse, not just in terms of gender but also background and culture. I am aware that research outlining barriers to engineering for women often points out the lack of visibility of women engineers and the lack of role models (World Economic Forum 2016). Participating in the competition was an opportunity to be both. And in addition to the evidence provided by my professional experience, at a personal level, I also believe that the worlds of engineering and the arts would benefit from greater integration. As engineering does not always provide an obvious avenue to exercise creativity, in some cases it can seem to be lacking in terms of allowing for individual expression. This

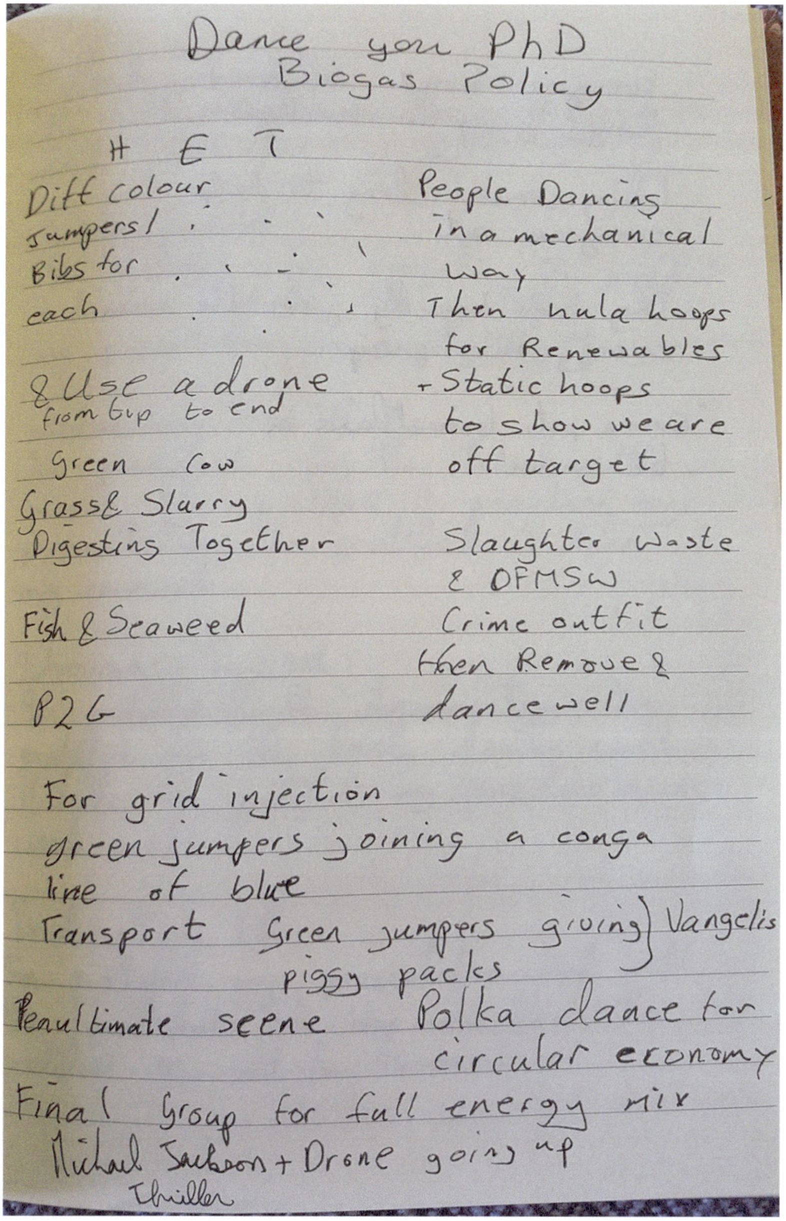

Image 6.1 Initial Draft – translating my PhD topic to dance

is at a time when creativity is needed more than ever, to address large-scale problems such as climate change and energy transitions. Aware of all these issues, the Environmental Research Institute, where the Biofuels and Bioenergy Research Group is based, has hosted artist Aoife Desmond from 2015 to 2016, who produced a 16 mm film that featured the ERI Lee Road building and researchers. The first public display of the work was in Crawford Art Gallery on Saturday 6th May 2017. The screening event and informal discussions around the artwork provided an important context for creative projects, as well as reassurance that my participation in the competition would have institutional support.

My Research Project

The purpose of this doctorate is to develop a roadmap for introducing green gas to Ireland. The Biofuels and Bioenergy Research Group has carried out extensive research on the resource potential for renewable gas from various feed stocks (Allen et al. 2016; Murphy et al. 2004; O'Shea et al. 2016). Key sources identified for Ireland include grass and slurry (O'Shea et al. 2017), organic waste (Browne and Murphy 2013) and woody biomass (Gallagher and Murphy 2013). Power to gas technologies are also being investigated (Vo et al. 2016). Once generated, the renewable gas can be used for combined heat and power, transported or injected directly into the gas grid. The project will have a strong emphasis on the renewable heat market and the potential to achieve upcoming mandatory EU renewable energy targets. Renewable gas can play an important role in decarbonising thermal energy as part of a combination of measures, considering dwelling or industry type, location, cost and emissions (SEAI 2017). Renewable gas is also known as 'green' gas. A green gas certification scheme to quantify emissions from green gas processes and the MWh traded will be investigated as part of my research (Long 2017). Also, lessons learned from other countries will be reviewed and key stakeholders engaged. The overall aim of the PhD is to propose a policy framework to incentivise feedstock cultivation and industrial process development and deliver thermal energy via green gas to appropriate markets in Ireland. And why dance it?

Science and Embodied Knowledge

Unlike traditional academic areas, the arts are an arena in which the body is central to the process of inquiry and indeed constitutes a mode of knowing. Embodiment can be defined as 'the integration of the physical or biological body and the phenomenal or experienced body, suggesting a seamless though often elusive matrix of body/mind worlds, a web that integrates thinking, being, doing and interacting within worlds' (Bresler 2004, p. 7). Whether using a neurophysiological, cognitive, phenomenological or sociological lens for its analysis, the role that embodiment plays in learning is now understood to be quite crucial. The body 'is the main forum for learning about how to do, think and believe, and practices as apparently diverse as belief and technologies are accordingly enacted and performed through the body in similar manners' (Stig Sorensen and Rebay-Salisbury 2012, p. 1). Dancing the

PhD then becomes a way in which not only is the research theme communicated in an entertaining and accessible way but also potentially a way in which learning about it takes place while choreographing and also participating in the dance. Indeed, it has been argued that effective instruction can build on recognising the embodied nature of scientific cognition and the role of metaphor in scientific thought and learning. Understanding abstract concepts is grounded, via metaphor, in knowledge derived from sensory and motor experiences arising from interaction with the physical world (Amin et al. 2016) In this case, the video is an interpretation of the title of my research project, namely, 'Renewable Gas Systems Modelling and Policy', which explains what renewable gas systems for Ireland are and how policy is relevant. I hope everyone who participated learned something about that. I elaborate on the significance of the dance in relation to the project in the section on 'Dancing It', below.

Although I have no formal training in dance, I greatly enjoy it and regard it as an important part of human expression. I became familiar with dance as a means to tell a story through attending the ballet performances that my mother brought me and my sisters to see when I was a child. My approach to the group dance scenes is informed by my participation in a flash mob for the short film Moore Street Masala, in which the producers asked participants to learn a dance that was uploaded to the internet for viewing, and on the day a professional dancer was hired to lead the crowd. I proceeded in the same manner, recruiting participants through an email sent to colleagues at the ERI.[2] I also started taking dance lessons in an informal way, taking part in open training sessions and workshops run by The Circus Factory in Cork and also hooping classes. The informal space of The Circus Factory was beneficial to develop ideas for the dance and practice skills.

Translating Research Project into Dance

The idea became a reality once I found a drone operator. My theory that there were people out there with hobby drones who would love to put them to good use was proved correct when I met Ciaran Usher at a social event. He worked in a technical role and was interested in the opportunity once I explained the idea to him. As an added bonus, he also turned out to have a background in television. From there I pitched the storyboard to my research team and supervisor in one of our regular meetings. Although I had given many presentations before, this was the first time I had a 'call-to-action' and genuinely cared about the result.

I outlined the requirements of the competition and also the 'story element'. I knew that researchers at the University of Alberta had modelled over 2000 stories and found that they all fit into 6 broad categories. I was familiar with this from general science reporting in the media, but one of the advantages of being an academic is that I can now look up reported studies on science direct. The six categories are

[2] See 'energy dance' at https://youtu.be/xnEwtpiH2YA

'rags to riches' (rise), 'tragedy' or 'riches to rags' (fall), 'man in a hole' (fall-rise), 'Icarus' (rise-fall), 'Cinderella' (rise-fall-rise) and 'Oedipus' (fall-rise-fall) (Reagan et al. 2016)[3]. As I was translating my research into a story, it was important for me that it fit into one of these patterns, which describe the emotional arc of the story (as opposed to the plot). My dance would be a Cinderella story. The reason I chose this format was to start with the excitement of possibility and then follow with the reality that what is possible is not in place. This then frames policy support as a saviour that can realise potential and end on a positive note again.

The group endorsed the project and agreed to participate. As mentioned above, the next step involved recruiting a chorus of dancers. I started recruiting 3 weeks in advance of the filming date as I wanted to have enough time to recruit but not so much that momentum would be lost. I also assigned hula hoops to people in the research group and close friends, so that they could learn hooping overhead. This would be representing the renewable energy targets.

Dancing It

Act 1: Framing the Problem

The opening scene features drone footage of the River Lee, which is a central feature of the city, before the drone rises up over the building and into the car park. The reason for this is that I wanted to pay homage to the opening scene of 'The Sound of Music' (Robert Wise 1965), where views of Salzburg feature in the opening credits. Also, as my research project is about Ireland, I wanted to make the video about Cork, and the Lee Fields are a popular walking spot. From the overhead drone footage, the next scene features the drone panning across the dancers in proportion. Electricity is represented by a dance well known as 'staying alive'. For heat, dancers started low and rose up with their arms waving, to represent rising heat and fire. Transport was represented by a driving motion.

The next scene features dancers standing in the car park, with some hula hooping a small hoop overhead. Initially, this scene was to be a pie chart with people to show the Irish energy use proportions of 20% electricity, 40% heat and 40% transport (Energy Modelling Group 2016). Given the renewable energy targets of 40%, 12% and 10%, respectively, I would have needed around 50 people for this scene. As I did not recruit this number, I instead used the car parking spaces to show proportions. The hula hoops represented the renewable energy supply targets within each energy type; as the targets are not yet reached, some hoops were dropped towards the end of the scene. The next transition scene is a photo of myself at my desk in the

[3] While these stories all come from a Western literary background, an added advantage of dancing science is that it can potentially reach audiences across cultures, with the steps, the rhythms and the choreography not primarily reliant on the written word.

research institute, asking what the overall aim of my thesis is: how do we meet our renewable energy targets for heat and transport?

Act 2: Feedstock to Energy

The feedstock to energy series of dances show the different potential sources of renewable gas in Ireland and how they can be converted to energy. I wanted my research group to play a starring role in this section and they kindly obliged. There were three elements to each dance, the 'before' to represent the feedstock, the 'process' dance and the 'after' or 'energy dance'. The energy dance was to be a common and hopefully recognisable theme throughout. I wanted it to look energetic, and as this would be the set-piece dance at the end, it also needed to be simple and easy to learn to encourage participation.

The first feedstock is residues represented by me and Tara Reddington. The punching-style dance represents the harm that residues such as food waste and slurry would cause if not processed correctly. The double turn represents the anaerobic digestion and then goes on to the energy dance. The double turn for anaerobic digestion represents two-phase continuously stirred tank reactors. This is the most common form of anaerobic digestion (Jerry D. Murphy and Thamsiriroj 2013). Grass was represented by a simple foxtrot, followed by double turn for anaerobic digestion and then the energy dance. I chose foxtrot for grass as marsh foxtail is a common grass variety in Ireland. I had thought of using a hula, but I felt that was not an Irish representation. The final feedstock for anaerobic digestion was seaweed. The partner dance represents the proposed cocultivation of fish farming and seaweed. The idea is that the seaweed can prevent eutrophication caused by fish farming, which would in turn lead to a higher seaweed, and therefore gas yield (Czyrnek-Delêtre et al. 2017). The partner dance ends before processing as the seaweed is harvested and transported for anaerobic digestion. Similar to the two previous scenes, the double turn represents anaerobic digestion before the energy dance to represent the transformation to energy.

The by-product of anaerobic digestion is called digestate; this can be spread on land as fertiliser and can replace slurry, which causes harmful emissions. In dance form this is represented by me and my sister performing a spreading action at the start before ending up in a 'tree' pose, representing growth as a result of fertiliser.

The next technology is gasification. The tree pose represents woody crops. Willow is proposed as a gasification feedstock for Ireland. There are a number of different types of gasification, and updraft gasification is suitable for willow, based on its water content (Quaak et al. 1999). For this reason, the hula hoop moves up my body to represent the process. After this I perform the by-now familiar energy dance.

The last technology represented is Power to Gas. The part I danced represents excess renewable electricity, which can be used for electrolysis. The turning movement represented the renewable element. Truc Vo and Richin Lin played water, with the arm movement representing the ripple of water, with Shane McDonagh as carbon dioxide, using a punching-style dance to represent the harm caused by this gas. I break through the water to split, and then the turning, shot overhead, represents the chemical reaction of hydrogen and carbon dioxide to form methane. Finally, Truc Vo and Shane McDonagh perform the energy dance to represent conversion to energy. Both are studying Power to Gas for their respective doctorates. As a follow-up activity, it will be interesting to find out to what extent embodying the metaphors for energy and gas behaviour that I devised to stage the dance also had a bearing upon their own research.

The circular economy is represented by the Irish Polka, a dance familiar to many Irish people, and often performed at weddings and social events. I chose it because it is a circular dance and also as a way of being distinctly Irish in an international competition. Renewable gas has important applications as part of the circular economy, as the resources are used and valued as much as possible (Wall et al. 2017).

Act 3: Uses of the Energy

The conga line, with people dressed in black to stand for fossil fuel, represented the natural gas grid. I was fortunate that there was good light shining on the hill down to the research centre. Renewable gas, which can be upgraded to the same standard as natural gas, was represented by the people who had been doing the feedstock to energy dances wearing green t-shirts. These were all part of the Biofuels and Bioenergy Research Group, including the Principal Investigator. The green represented renewable gas. For transport I chose a hula hoop dance to 'The Wheels on the Bus', a popular nursery rhyme. The hula hoop represented the turning motion of wheels.

Act 4: Energy Dance

The framing of the story, followed by the potential energy sources and uses, was meant to convey the 'rise' stage of the story. The 'fall' or the emotional low point was the next scene, where the potential for renewable gas has not yet been reached and we are sitting on the sidelines while everyone else is having fun. We look bored and are not participating. The second 'rise' of the story occurs when policy, played by Serena Lee O'Sullivan, comes along, in a flamboyant way, to rescue us. Policy gives some of us a hand up and the others get up of their own accord, as the industry is established. The silent movie text reads 'Nobody puts renewable gas in the corner', referencing the famous line from 'Dirty Dancing' (Emile Ardolino 1987): 'nobody puts Baby in the corner'. Without policy support, renewable gas is in the corner. In the final scene, both renewables and fossil fuels are all part of the energy system, performing the energy dance in an ongoing loop.

Choreographing the Research/Dance

For the energy dance, I wanted a simple dance that would be easy to learn but that would also take up space and look energetic. It also needed to be manageable in a group, so that participants would not be bumping into each other. I recorded a video of myself dancing to the music in the back garden and mirrored it, so you could learn it facing a screen. To record the consent of participants who responded to the email that I mentioned above, I created a google form as a sign-up and also had manual sign-up on the day.

Although several colleagues initially signed up, some were put off by the dance, which they felt was fast and complicated. To address this I filmed a second step-by-step version. To cast a wider net, I also sent the recruitment email, giving details of the competition and a link to the dance to the entire UCC staff email list of 4000 people. I got no volunteers from this. I also reached out to dance groups through any contacts I had. This did not yield any participants either. Most were a result of direct conversations. In the end the energy dance participants were approximately half researchers from the ERI and half family and friends.

It was a requirement of the competition that I would be dancing in the video. As producer, I also wanted a starring role. I assigned tasks according to the personalities of the people taking part, in terms of how willing and comfortable they were to take on prominent roles on camera. I knew Michelle Healy as someone with high energy and enthusiasm, so I asked her to lead the conga line and also to lead the final energy dance off-camera. Knowing that I would not be in a position to greet people fully as they arrived, they were sent to her for a rehearsal before filming started. This served two purposes: to get to know the dance if participants had not learned it fully and also to mitigate the awkwardness that they might feel when initially arriving at the event. Many participants commented afterwards that they thought Michelle was an aerobics instructor or that it was what she did for a living.

From my own research group, some were enthusiastic about the project, but not willing to participate on camera. I assigned Karthik Rajendran the role of first assistant director. His role was to greet people on arrival, to assign energy type for the opening scene and to assist with organising the dance group. He was also responsible for the clipboard, i.e. my storyboard, in order of shooting sequence. Karthik Rajendran had an interest in photography and a good camera and offered to document the day. This resulted in some lovely 'behind the scenes photos' that feature in this chapter. Dawn O'Sullivan had also agreed to assist, but not on camera. I assigned her the role of second assistant director, playing the music according to a list I had prepared to match the storyboard and also helping with organising the large group. Síle Griffin was my phone camera person, who provided insight on the quality of scenes and also had great patience, and finally Ciaran Usher was the drone camera person. Apart from general briefings, I found that I worked through my team, who then managed the large group.

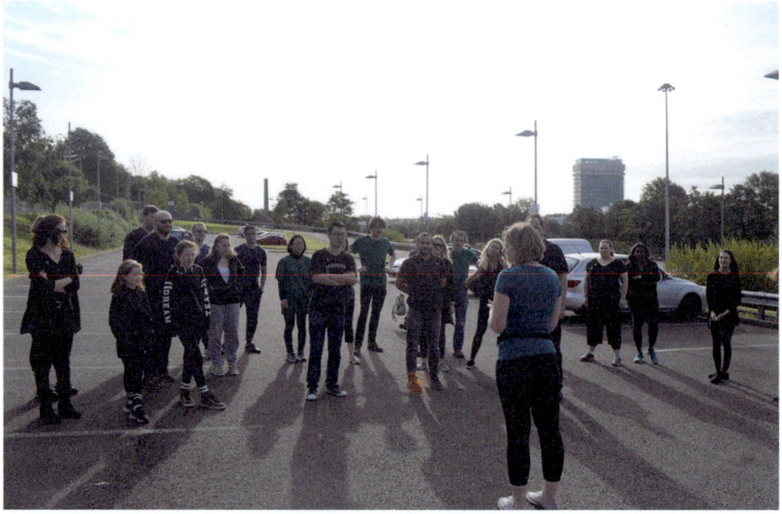

Image 6.2 Briefing participants on the shoot requirements and research

I asked people to arrive between 8.45 and 9 and advised that we would not be there longer than 11 (see Image 6.2). Before filming I started with a briefing. The group was not all familiar with the research and that was one of the aims of the day. Before each group scene, I explained what the scene would represent.

I started filming with the conga line scene, as it was less dependent on a large crowd compared to other group scenes, and it would wake the group up and get everyone into a dancing mood. This was inspired by a humanitarian UN Civil-Military Coordination training I had participated in. Each morning there were learning activities meant to reinforce the material covered the day before. One day the game was a conga line. While initially I despaired of the exercise, by the end I was fully enthusiastic. I hoped to achieve the same result by starting with a conga line and it did work (see Image 6.3). The following scene was where renewable gas is on the sidelines and rescued by policy. Policy was played wonderfully by Serena Lee O'Sullivan, a trained dancer and model. This was the set-piece dance and final scene, and at that point, all dancers were fully engaged and enthusiastic.

The final group scene was the opening scenes, representing the energy types and proportions. By this time everyone had arrived. Shooting this scene was led by drone operator Ciaran Usher. This scene also took the longest as it was difficult to synchronise the timing of the drone speed with that of the dance. When the group scenes were finished filming at 11, tea/coffee prepared by David Wall was served in the building reception. After a short break, we continued filming the smaller group scenes and then finally the solo scenes. In total, filming took from 8.30 a.m. to 12.30 p.m. (see Image 6.4). Editing took approximately two weekends.

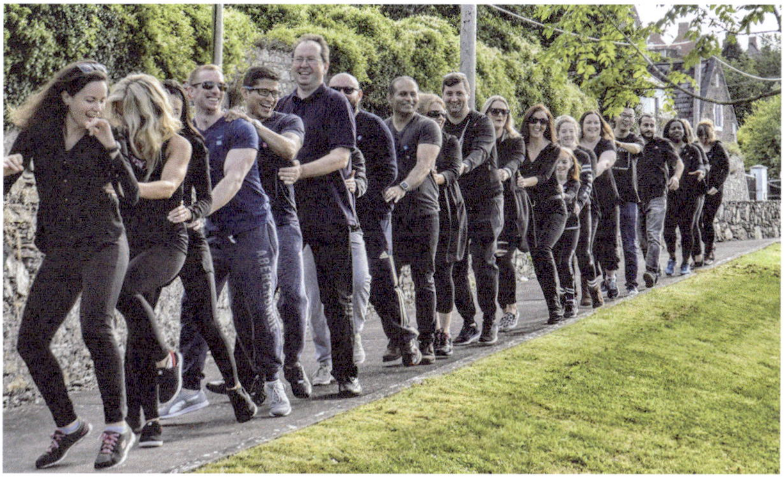

Image 6.3 Conga line – first scene of the day to energise the group

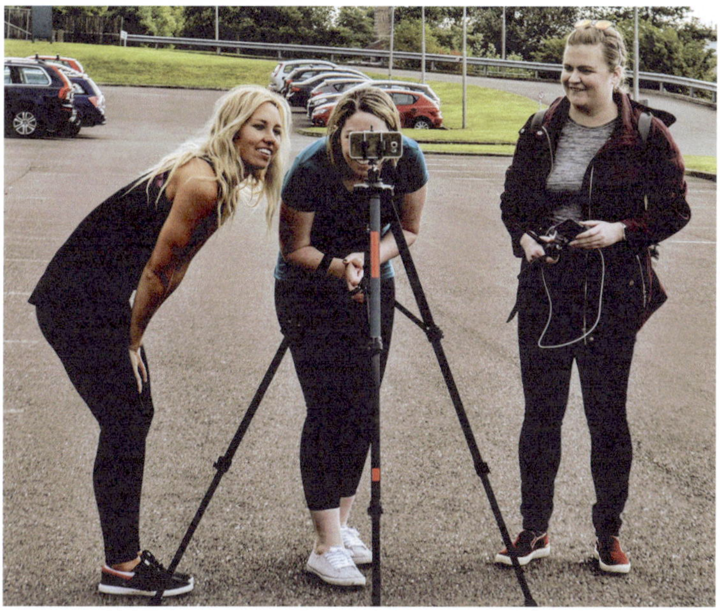

Image 6.4 Reviewing the film footage with Serena Lee O' Sullivan and Síle Griffin

Challenges Faced

The first challenge for me was to be able to overcome some personality traits. While I do not consider myself shy, I am not particularly outgoing either. It was a large undertaking to ask everyone I saw for 3 weeks whether they might be interested in my 'Dance your PhD' project. Moreover, I got quite ill while still in the recruiting process. Being far out of my comfort zone may have been a contributing factor, because although I was confident I would get enough participants, I was also very aware that every participant was hard won. Early on I decided to focus my gratitude on those who agreed to participate rather than be disappointed with those who committed and then declined or who were not interested. This was important to maintain mental momentum and enthusiasm for the project.

I also learned that very few people fill out an online form, especially if asked over email. Every form sign-up was a result of direct conversations. With hindsight, I should have been more aware of this from my time working in data collection, so it was an interesting refresher. Moreover, following on from email requests, the project demonstrated the importance of networks. All participants were from my own network or knew people joining in. There was no response from the general request or reaching out to dance groups. A good idea is not enough and there needs to be trust. This is an important lesson learned to apply to building links between disciplines, on either a professional or personal basis.

Having respect for the participants was important, ensuring that they were comfortable, starting on time, providing access to toilets, keeping people in the loop as to the 'why' with briefings and also ending on time. I managed to achieve this, which was important to the success of the day. Before the end of the filming day, I sent a thank you email to all participants, with a 36 s trailer. This served two purposes: first, it was a tangible outcome to show for the day, and second, it was an indication as to when the video would be ready. I credited everyone that participated and also dance instructors that I had met along the way. Finally, after the video was done and uploaded, I followed up with thank you cards, acknowledging the contribution that each person made.

I also gained some interdisciplinary training in the sense that I became familiar with the process and practicalities of making a video. From a directing point of view, I learned about the importance of management structures, having a team and trusting them in their role so that the day ran smoothly. It was tempting to try and assume full control of the day, but I would not have been able to manage everything at once. Those who volunteered to be behind the scenes were highly valuable to the success of the day, so team work was and is absolutely crucial.

Understanding social media platforms and networks was important for spreading the word, and participating in the competition provided an invaluable opportunity to become engaged with these. In the social media training mentioned previously, we were advised that it is not a good idea to post a YouTube video on Facebook, as the companies are rivals and Facebook would bury the video in newsfeeds. For this reason the video was uploaded directly to Facebook on the MaREI Facebook page and on my own YouTube account. Most of the views were on Facebook. At the time of writing, momentum is slowing down, and without advertising or promotion in traditional media, the video has approximately 2500 views on Facebook and 500 on YouTube, over 3000 views in total and counting.

Overall, the video has been well received. It was seen as something fun and informative. From informal discussions with people who participated and friends, they now understand my research from a general point of view. To quote one participant 'Now I finally know what you're doing!'

Conclusions

While this video was shot out of sequence, to manage time and the availability of participants, the dance could be reformatted into a dance workshop if done in sequence. This would continue the storytelling element and engage people who may not have a primary interest in science.

Finding a way to build relationships between disciplines is important for trust and collaboration. This project depended very much on my personal contacts, and it showed how weak relationships did not have a positive outcome for involvement in the project.

There is no doubt that the project has been beneficial for me as an engineer, to improve my ability to explain what I am doing to a non-technical audience, both at the event, through the choreography, and writing the script for the video. As a first year researcher, I spend most of my time perpetually confused as I am reading about my topic. Working on the video gave me an opportunity to focus on the big picture as well as the detail of my work. Moreover, to follow up on the insights I gained this year, I intend to continue designing dances for my PhD every year regardless of the competition, to integrate the knowledge I gain from the construction of metaphors and the physical acting out that the process involves.

I have been at dance performances which tell either no story or a terrible story. Scientific research essentially tells a story. As Hans U Fuchs has demonstrated, narrative framing is an essential component of scientific modelling in physics and beyond (Fuchs 2017). Translating science stories to dance performance can have a benefit for society and contribute to the cultural fabric of the city community as well. Science is full of potential stories, waiting to be told.

Acknowledgements This material is based upon works supported by the Science Foundation Ireland (SFI) under Grant No. 12/RC/2302. The researchers of the Biofuels and Bioenergy Research Group are employed by the SFI Marine and Renewable Energy Research Institute (MaREI). Gas Networks Ireland (GNI) co-founded the work through funding of the Gas Innovation Group. My research is supervised by Professor Jerry Murphy and Professor Brian Ó Gallachóir.

References

Amin TG, Jessopsson F, Haglund J (eds) (2016) Conceptual metaphor and embodied cognition in science learning. Routledge, London/New York

Bresler L (2004) Prelude. In: Bresler L (ed) Knowing bodies, moving minds: towards embodied teaching and learning. Kluwer, Dordrecth

Fuchs HU (2017) From stories to scientific models and back: narrative framing in modern macroscopic physics. In: Amin TG, Jeppsson F, Haglund J (eds) Conceptual metaphor and embodied cognition in science learning. Routledge, London/New York
Stig Sorensen ML, Rebay-Salisbury K (2012) Embodied knowledge: reflections on belief and technology. In: Stig Sorensen ML (ed) Embodied knowledge: historical perspectives on belief and technology. Oxbow, Oxford, pp 1–10
Allen E, Wall DM, Herrmann C, Murphy JD (2016) A detailed assessment of resource of biomethane from first, second and third generation substrates. Renew Energy 87:656–665
Amin TG, Jessopsson F, Haglund J (eds) (2016) Conceptual metaphor and embodied cognition in science learning. Routledge, London/New York
Bresler L (2004) Prelude. In: Bresler L (ed) Knowing bodies, moving minds: towards embodied teaching and learning. Kluwer, Dordrecth
Browne JD, Murphy JD (2013) Assessment of the resource associated with biomethane from food waste. Appl Energy 104:170–177
Czyrnek-Delêtre MM, Rocca S, Agostini A, Giuntoli J, Murphy JD (2017) Life cycle assessment of seaweed biomethane, generated from seaweed sourced from integrated multi-trophic aquaculture in temperate oceanic climates. Appl Energy 196:34
Energy Modelling Group, S (2016) Ireland's energy targets
Fuchs HU (2017) From stories to scientific models and back: narrative framing in modern macroscopic physics. In: Amin TG, Jeppsson F, Haglund J (eds) Conceptual metaphor and embodied cognition in science learning. Routledge, London/New York
Gallagher C, Murphy JD (2013) What is the realistic potential for biomethane produced through gasification of indigenous willow or imported wood chip to meet renewable energy heat targets? Appl Energy 108:158–167
Long A (2017) greengascert.ie. Retrieved September 18, 2017, from http://www.greengascert.ie/
Murphy JD, Thamsiriroj T (2013) Fundamental science and engineering of the anaerobic digestion process for biogas production. In: The biogas handbook. Woodhead Publishing Online, Philadelphia, pp 104–130
Murphy JD, McKeogh E, Kiely G (2004) Technical/economic/environmental analysis of biogas utilisation. Appl Energy 77(4):407–427
O'Shea R, Wall D, Kilgallon I, Murphy JD (2016) Assessment of the impact of incentives and of scale on the build order and location of biomethane facilities and the feedstock they utilise. Appl Energy 182:394–408
O'Shea R, Wall DM, Kilgallon I, Browne JD, Murphy JD (2017) Assessing the total theoretical, and financially viable, resource of biomethane for injection to a natural gas network in a region. Appl Energy 188:237–256
Quaak P, Knoef H, Stassen H (1999) Energy from biomass: a review of combustion and gasification technologies. World Bank Tech Pap 422:1–78
Reagan AJ, Mitchell L, Kiley D, Danforth CM, Dodds PS (2016) The emotional arcs of stories are dominated by six basic shapes. EPJ Data Sci 5(1):31
Science (2017) Dance your PhD official rules. Retrieved September 15, 2017, from http://www.sciencemag.org/projects/dance-your-phd/official-rules
SEAI (2017) Assessment of cost and benefits of biogas and biomethane in Ireland
Stig Sorensen ML, Rebay-Salisbury K (2012) Embodied knowledge: reflections on belief and technology. In: Stig Sorensen ML (ed) Embodied knowledge: historical perspectives on belief and technology. Oxbow, Oxford, pp 1–10
Vo TTQ, Xia A, Wall DM, Murphy JD (2016) Use of surplus wind electricity in Ireland to produce compressed renewable gaseous transport fuel through biological power to gas systems. Renew Energy 105:495–504
Wall DM, McDonagh S, Murphy JD (2017) Cascading biomethane energy systems for sustainable green gas production in a circular economy. Bioresour Technol 243:1207
World Economic Forum (2016) The industry gender gap
Wise R (1965) The sound of music. 20th Century Fox
Ardolino E (1987) Dirty dancing. Great American Films Limited Partnership

Chapter 7
Collaborative Journeys in Art and Biology

Anna Dumitriu

Abstract This chapter describes the artistic journey of internationally renowned BioArtist Anna Dumitriu and her artistic process of working on long term laboratory based artistic residencies and collaborations. Dumitriu works at the cutting edge of science, drawing threads across time to explore the history and future of technology, biomedical science, healthcare and ethics. From her fine art training to her work with CRISPR gene editing techniques and antibiotic resistance, her work fuses traditional media such as textiles, sculpture and traditional crafts with new media such as pathogenic bacteria and DNA, to create emotionally affecting objects and installations that are exhibited around the world.

Keywords Art · Science · BioArt · Public engagement in science · Collaboration · Biomedicine

As a UK-based artist, I have worked for many years by creating my work embedded in biomedical, scientific and technology-based settings, travelling around the world to explore the different areas of research taking place and thinking about the personal, cultural and societal issues that are implied by the research in those contexts and beyond.

I was originally trained in fine art, but now I easily spend as much time in the laboratory as in the studio, and the materials I use range from traditional art and craft media to modified DNA or robot motors. My first collaborations date back to the turn of the twenty-first century, which sounds a very long time ago, and I suppose it is as I have seen significant changes already. The themes I explore now continue to build on that early work I did looking at genetics, microbiology and computer science, which has given me a fascinating perspective on the development of those fields and the massive impact of new research tools and technologies. The artworks I produce through my collaborative projects are exhibited in galleries and

A. Dumitriu (✉)
Brighton and Sussex Medical School, School of Computer Science, The University of Hertfordshire, Hatfield, UK
e-mail: annadumitriu@hotmail.com

© Springer International Publishing AG, part of Springer Nature 2019
A. de la Garza, C. Travis (eds.), *The STEAM Revolution*,
https://doi.org/10.1007/978-3-319-89818-6_7

museums across the world, as well as in nontraditional settings such as hospitals, and aim to aesthetically reflect on new research in such a way that it helps to create a dialogue across diverse audiences through an aesthetic perspective.

Over the years I have created many collaborative projects in laboratory or healthcare settings, and my working relationships with researchers tend to keep going, sometimes working together intensely and sometimes having quiet periods as other things take over or as we inevitably wait until it is possible to secure funding to develop our ideas, which can sometimes take a while.

There has been a kind of logical progression in my work as I explore my areas of interest, but there has also been an element of serendipity which has been developed through meeting new people and making connections. As I create work and build on my knowledge, I have reached the stage where I am also approached by scientists as much as having to seek out scientific collaborators. From starting with just a handful of collaborators, I now have quite a few, and we keep in touch over social meetings, cups of coffee or conference meetings, as well as at exhibitions of our collaborative works or when we run collaborative workshops together. Sometimes my collaborators also start to collaborate on science projects, which are interesting to see.

My artistic journey into the scientific world began with an interest in cellular forms and the theme of immortality during my postgraduate degree in fine art at the University of Brighton. Based on that work, I received an invitation to undertake a short residency in the Clinical Genetics Laboratory at St. George's Hospital in London around the year 2000 where I started to look at cell division and cell death under the microscope. This was my first insight into genetics, which was a field I knew little about at the time and that has changed incredibly since I started. Leading on from this, I became fascinated in learning more and peeling away at the layers of knowledge and building on my experience along the way. In a sense my work has evolved along with new discoveries in science as described below.

About 15 years ago, I became fixated on an interest in 'normal flora', which was the common name at the time for the ubiquitous micro-organisms that co-exist with us, not particularly the disease-causing ones but the ones that are just there. These were described at the time to me as being 'of no commercial or medical interest', and that fascinated me.

I came to understand later that this was because we did not have the tools to study them. But I saw it as a philosophical question at the time: Where do we draw the limits of what we can study? And if it is of no commercial or medical interest, then maybe it is of artistic interest.

I decided to try to work with a scientist to explore this, and I took a gamble and sent an email to a general contact address at the newly opened Brighton and Sussex Medical School and asked if they could forward my message on to the academic researchers there. I talked about my past work and my fascination for bacteria; I sent some links and images of what I had done before. Luckily, I got a reply from Professor Helen Smith who liked my work and was interested in collaborating, but she was not a microbiologist. Instead she specialised in allergy and was the Chair of Primary Care. Helen invited me to work with her and apply for a Wellcome Trust Pulse Award, which was a fund to enable public engagement projects across art and biomedicine to work with young people. We were successful in our funding application, and we started 'The Sensitive Project' which explored the impact of allergy on the lives of young people working with Varndean School, a large secondary school in Brighton, UK.

'The Sensitive Project' (Smith 2007) explored notions around the hygiene hypothesis, the idea—which has significantly evolved since the time of the project—that the prevalence of allergy is caused by our lack of exposure to dirt. The project also looked at how the immune system works in relation to an allergic reaction. We worked in a participatory way with the teenagers, and our artistic outcomes included textile artworks and installations, such as an intervention in the form a painting on the school playground which disappeared each time it rained.

The project crossed boundaries between art and science enabling a greater understanding of allergy through scientific research fused with creative activities such as live art, video and installation art. As part of the final event, the whole school was turned into a large-scale artwork for one evening, and parents, medics and many members of the art community attended the event. The project linked art and science, specifically installation, intervention and performance art, to allergy. I fully believe that to express any concept or idea, both art and science are needed, and through art we can create a synthesis between our emotional responses to the world and the scientific analysis of it, which of course is not to say that science is not emotionally driven, too. Of course it is and much more than is spoken about in the media. The choices that scientists make, the passionate ones at least, can be very much aesthetically driven.

The concept of 'The Sensitive Project' was interesting as allergies pretty much affect everyone on a personal level in some way, if not at firsthand, then through friends and family. Everyone knows someone with an allergy it seems. But there is a huge amount of misunderstanding about how allergic reactions actually happen, what effects they have physically and what the best ways of dealing with them are, not to mention the restrictions they place on people's lives. We chose to work with year 9 and 10 pupils on this because as an age group, they are frequently most affected; from an artistic point of view this was a conceptual choice and added a deeper layer of meaning to the final artworks. From a scientific point of view, the work was about creating an understanding of the subject for the students but also about feeding back into patient management strategies some of what was learned from actually working with the students.

The gifted and talented science students we worked with at the Varndean School made constant connections and suggestions to the development of all the artworks, and the students learned new ways of expressing their ideas and thinking about how art can help them do that, especially when it exists outside the traditional gallery space in their everyday lives. We worked with installation, performance, interventions and even a GPS drawing, which utilised a newly accessible technology at the time, to create a map of allergens in the school grounds. It was inspiring to work with a scientist and think about the emotional impact of allergy as well as the mechanisms behind it. The project even went as far as teaching the young people how to give adrenaline injections to people affected by anaphylactic shock caused by an allergic reaction, a useful skill given the recent death of a teenager, Nasar Ahmed, at a school in the UK from an allergic reaction.[2]

Part of the way through the project, Professor Smith introduced me to Dr. John Paul, a microbiologist who was willing to meet me and possibly collaborate. The meeting went well; we started to work together and have collaborated ever since. Dr. John Paul has been my most important collaborator, for 15 years so far, and we are still working together and developing projects; we've been emailing back and forth as I write this now.

It was through John that my practice evolved to work with infectious diseases, an area that artists seem to shy away from or perhaps cannot gain access to work with but I consider incredibly important, especially in terms of working with the wider public. The issue of antibiotic resistance in infectious diseases is probably one of the most significant threats we face, and public engagement can help hugely in cutting down antibiotic misuse whilst allowing people to ask questions that matter to them.

Our 'Normal Flora Project' (Dumitriu 2017) started in 2004, focussing on these organisms that science didn't bother to study at the time, and we started to explore the ubiquitous bacteria, moulds and yeasts that form an absolutely key part of the complex ecosystems we live constantly with, from our bodies, our homes and our everyday world and, of course, the wider planet. Specifically, I started to map the bacteria in my own home.

The word 'bacteria' is synonymous with dirt; the normal reaction to the suggestion that something is covered with bacteria is one of disgust, but in fact almost everything is covered with bacteria. The project offered a physical embodiment of the interconnectedness of life through a deep examination of the rarely noticed everyday world of microbial life we co-exist with. I like to think of the project as not asking 'how clean is your house?' but instead 'how sublime is your ecosystem'.

It has been fascinating to observe how research has developed in the intervening years as tools such as sequencing technologies have become far more accessible and we are now starting to understand a little more about these bacteria which live alongside us. I have observed as these previously disregarded organisms have been rebranded as 'the microbiome' and seen the beginning of a kind of gold rush of bioprospecting for novel micro-organisms that might contain things that are useful to us, such as enzymes for washing powders, or ways of creating new antibiotics.

When John began working with the Modernising Medical Microbiology project (Medical Microbiology 2017), I became enthralled in understanding the bacteria they were studying. The Modernising Medical Microbiology project led by the University of Oxford, Nuffield Centre for Clinical Medicine, looks at the changing face of medical microbiology, in light of the possibilities of near real-time whole-genome mapping of bacteria and developments in bioinformatics. Their research focussed on clinically important and newsworthy organisms including *Mycobacterium tuberculosis*, *Staphylococcus aureus* and *Clostridium difficile*. I'm still obsessed with these organisms, which have such a massive impact on patients, and continue to be amazed as science constantly seems to be developing new understandings of them whilst simultaneously realising that we have so much more to learn.

John insisted that I be trained in microbiology, though not really in any official sense. He taught me in the lab, and I did all my experiments myself following proper procedures and fully documenting all the work and results in a lab book. He also taught me to make my observations in the way I would understand them when I looked back over the book later on, so my artistic reflections were given as much weight as scientific results.

Much of my work involves participatory elements, and John and I have also worked together to develop do-it-yourself microbiology activities that could be done safely with members of the public, such as in my 'Infective Textiles' project, which involved exhibiting a 4 m agar plate grown with environmental bacteria. Our collaboration eventually evolved into my role as an artist in residence with the Modernising Medical Microbiology project and my ongoing artistic research into antibiotic resistance, MRSA, tuberculosis and the gut microbiome.

I received a 1-year Leverhulme Trust Artist in Residence Award in 2011 to support my residency at the Nuffield Department of Clinical Medicine at the University of Oxford, and my work also takes place at partner centres in Brighton, Leeds and Birmingham. Since 2012 I have continued in my artist-in-residence role on an ongoing basis to further develop my work, sourcing funding where I can on an ad hoc basis. The aim of my work there is to artistically explore the aesthetic and cultural implications of the research and how it matters to ordinary people. I aim to communicate the impact of new technologies in microbiology and how they will improve understanding, diagnosis, treatment and control of infections as well as to think critically about them and bring an artistic voice into the lab.

Nowadays it is becoming possible to sequence whole bacterial genomes, make comparisons and even see whether bacteria carry genes for antibiotic resistance, and this will revolutionise the treatment of infectious diseases, worldwide. Although it's an expensive technology, bacteria and viruses do not take any notice of borders, and we need to help to combat disease wherever it arises. Bacteria are the most diverse and successful life forms on Earth, and with this technology, we can know them a little better.

My "MRSA Quilt" was the first major project I created with the Modernising Medical Microbiology Project, and I worked with Dr. John Paul and Dr. James Price to create the work based on a conversation I had with James on my first visit to his lab. He told me about the diagnostic chromogenic agar jelly they used to grow the bacteria for testing, a kind of seaweed-based growth medium that makes the bacteria grow in a different colour. He worked with an agar jelly that made MRSA, the antibiotic-resistant superbug form of *Staphylococcus aureus* grown 'a lovely denim blue'. I had the idea to embed cotton calico in the chromogenic agar and to see if we could actually keep the blue colour if the cloth was sterilised in an autoclave. It worked, and I stitched hundreds of MRSA dyed squares of calico together to make a kind of storytelling quilt.

The patterns on the calico squares were created using different tools and techniques in the treatment and diagnosis of MRSA. The patterns include stripes and polka dots created using antibiotic susceptibility tests, such as vancomycin susceptibility discs and cefoxitin strips, and embroideries made using natural antibiotic dyed threads. Quilts are a traditional way of passing down stories, and the piece also became a discussion tool to facilitate dialogue between the wider public and the scientific research team. We ran workshops together at a range of locations including a large-scale drop-in activity at the V&A Museum in London, where around 200 visitors came and made their own MRSA Quilt squares. Visitors worked alongside the scientists and I and could chat and ask questions about the research and the healthcare issue in a casual and relaxed way. We added the squares and different natural and clinical antibiotics in the museum and added the MRSA back in the lab later. We took photos of the results and put them online for people to see, and then we sterilised the public contributions, and I used them to make other works, such as 'MSSA to MRSA: Becoming Resistant'.

As the research developed, I started to shadow the work of Kevin Cole, a biomedical scientist focussed on whole-genome sequencing of bacteria. It took me a few years to get the courage to do it, but I wanted to explore this important technology, and I decided that in order to properly understand how it worked, I had to learn how to do it myself, the entire process, which almost everyone thought was perverse as it's so fiddly and complicated and has so many steps from preparing DNA for sequencing to actually doing the bioinformatics and computational side. There is a huge amount of data involved. It's one of those subjects frequently cited in discussions on 'big data'. I decided to use the results and the media produced on my journey through Modernising Medical Microbiology's research to make a new artwork. Arts Council England funded the project.

I found that I was colonised by *Staphylococcus aureus* bacteria as indeed are at least 30% of us. So I tested and sequenced the genome of that *Staphylococcus aureus* that lives on my body. The spa gene (surface protein A gene) of the *Staphylococcus aureus* bacteria found to be living at the front of my nose was type t015, which corresponds to sequence type ST45 and is not the most common type seen in hospitals. However, my *Staphylococcus aureus* is actually resistant to penicillin. By making my colonisation with this feared organism explicit, I tried to demonstrate the gap between the media's presentation of this bug and the scientific reality. The point being we are colonised by a huge number of bacteria, many of which are integral to our well-being. Even my *Staphylococcus aureus* may play a part in my health and well-being. The science behind this work is revealing that there is a far greater story to be told than the 'dirty hospitals' rhetoric of the press, and our work offered a way into the story for many people who would otherwise have no access to it.

The resulting artworks took the form of two bio-digital pieces and collaborated with digital artist Alex May to help me work with the data I had produced. A video mapped a dress that was made from calico grown with the *Staphylococcus aureus* from my body, MRSA and VRSA which is also resistant to the antibiotic vancomycin. Projected onto it was the raw light output of the whole-genome sequencer as it processed my bacteria using fluorescent molecules. The other artwork was a virtual reality exploration of the vastness and complexity of the data of just this one tiny organism.

One of the things Kevin Cole and I always talk about is the 'boring' bits of doing science. It has become so important that we have even talked about it in lectures. What I mean by this is in working with whole-genome sequencing or synthetic biology, there is an enormous amount of pipetting of tiny amounts of liquid into other tiny amounts of liquid and waiting for PCR machines to run. The pipetting takes concentration and a very methodical approach to make sure you do not pipette the wrong liquid into the wrong tube, but waiting for the PCR is just waiting, and so that's when we chat and when ideas come. We talk about the work in such a way that we would never do if we were having a meeting and being organised. We literally chat to fill up the boring minutes, and for me I think that's kind of when the magic happens and I get new ideas.

Learning that my *Staphylococcus aureus* is resistant to penicillin was fascinating to me. Around 80% of *Staphylococcus aureus* are now resistant so it's not unusual. I've been working with the Museum of the History of Science in Oxford who held a major historical exhibition called 'Back from the Dead' to commemorate the seventy-fifth anniversary of the first use of penicillin in a human patient, the start of the antibiotic era. I started to wonder if it would be possible to use modern techniques from synthetic biology to 'repair' my *Staphylococcus aureus* back to its pre-1941 state. I started to look at the newly discovered technique of CRISPR (Clustered Regularly Interspaced Short Palindromic Repeats), which allows DNA to be cut and pasted into the genome of an organism, and wondered if I could edit out my penicillin antibiotic resistance gene.

In 2017 I started a Future Emerging Art and Technology (FEAT)[1] residency (Oxford Museum of the History of Science 2017) in collaboration with the EU Future Emerging Technology (FET) project MRG-Grammar (FEAT 2017). A key part of my residency was working with Dr. Sarah Goldberg and Dr. Roee Amit at the Synthetic Biology Laboratory for the Decipherment of Genetic Codes at the Technion in Israel. They are trying to understand how genes are regulated in bacteria and are working with CRISPR gene editing. It so far hasn't been possible to edit out the penicillin resistance gene from my *Staphylococcus aureus*, but that is a future aim. *Staphylococcus aureus* is what is known as a biosafety level two organism as it has the potential to cause disease in humans and therefore requires extra careful handling, and many synthetic biology labs are biosafety level one.

What Sarah and I did manage to do however was to take a biosafety level one *E. coli* bacterium and edit its genome to remove an ampicillin resistance gene, and this is not a trivial experiment in the way we did it. The final piece I made drew threads across time and explored ideas around 'control' and 'misuse'. The finished work was first exhibited at LifeSpace Dundee in April 2017 (MRG-Grammar 2017) and takes the form of an altered antique wartime women's suit marked with the British Board of Trade's utility logo CC41, which stands for 'Controlled Commodity 1941' meaning that the use of materials was deemed to meet the government's wartime

[1] FEAT is an initiative of eutema GmbH (AT), Stichting Waag Society (NL) and youris.com (BE). It has received funding from the European Union's Horizon 2020 research and innovation programme under grant agreement No. 686527 (H2020-FETOPEN-2015-CSA).

austerity regulations. The holes and stains in the suit have been patched and embroidered with silk patterned with *E. coli* bacteria grown using a dye-containing growth medium, forming pigmented colonies or spots. The genomes of these *E. coli* bacteria were edited using CRISPR to remove an ampicillin antibiotic resistance gene and repaired using a technique called homologous recombination to scarlessly patch the break with a fragment of DNA encoding the WWII slogan 'Make Do and Mend', which encouraged housewives to repair their clothes during the wartime rationing period. The phrase was converted into ASCII code and then to base 4 and into the DNA bases ATGC. Ampicillin is part of the penicillin group of beta-lactam antibiotics, so in a way it is conceptually and poetically true to say that, with this artistic genomic edit, we have used today's latest technology to 'mend' the organism back to its pre-1941, pre-antibiotic era state, but scientifically it is far more complex in that they have used a lab strain of TOP10 *E. coli* that is very well characterised and has had many other modifications, so it will never really be the same as it was in 1941.

Through one of the team members of the MRG-Grammar project, I was introduced to another EU FET project called BeyondSeq, and it was suggested that I develop a collaboration with Dr. Rob Neely at the University of Birmingham to explore how he works with fluorescent molecules to visualise DNA. We managed to secure a Leverhulme Artist in Residence award to fund the work, and so now I am currently trying to develop ways of visualising parts of the genome of my special Make Do & Mend strain of *E. coli*. In a way this may again lead to scientific as well as artistic developments and collaborations as it's such a new field (LifeSpace Dundee 2017).

For me collaboration has become a journey, and somehow I feel my work is weaving threads across time from ancient tuberculosis treatments to the cutting edge of synthetic biology and sequencing technologies and drawing together teams of people to bring a range of disciplines and perspectives to bear on a common interest. It is about communication between disciplines, with the wider public, and it is important for future generations who will be impacted by these technologies for a very long time.

References

Anna Dumitriu: Bioart & Bacteria (2017) Artist's website, Anna Dumitriu, Brighton. Viewed 12/03/2017. http://www.normalflora.co.uk

Arts meets science to promote awareness of allergy in a school, Smith, Helen E. World Allergy Organization Journal, November 2007

FEAT (2017) Project website, Eutema, Vienna. Viewed 12/03/2017. http://www.featart.eu

Introducing Artist in Residence Anna Dumitriu (2017) Article on art residency, University of Birmingham, Birmingham. Viewed 12/03/2017. http://www.birmingham.ac.uk/research/activity/sci-phy/news/Introducing-Artist-in-Residence-Anna-Dumitriu.aspx

LifeSpace Dundee (2017) Gallery website, University of Dundee, Dundee. Viewed 12/03/2017. http://lifespace.dundee.ac.uk

Modernising Medical Microbiology (2017) Project website, University of Oxford, Oxford. Viewed 12/03/2017. http://www.modmedmicro.ac.uk
MRG-Grammar (2017) Project website, Technion, Haifa. Viewed 12/03/2017. https://www.mrg-grammar.eu
Oxford Museum of the History of Science (2017) Museum website featuring 'Back from the Dead' exhibition, Oxford museum of the history of science, Oxford. Viewed 12/03/2017. http://www.mhs.ox.ac.uk/exhibits/back-dead-demystifying-antibiotics
See Gayle D (2017) Mother of boy who died from allergic reaction to dinner criticises the school. The guardian. https://www.theguardian.com/uk-news/2017/may/12/mother-of-boy-who-died-from-allergic-reaction-to-dinner-criticises-school-nasar-ahmed Accessed 8 Aug 2017
The Normal Flora Project (2017) Project website, Anna Dumitriu, Brighton. Viewed 30/05/2017

Chapter 8
From STEM to STEAM at the Beautiful Midden Field School: An Artist/Educator Perspective

Anita McKeown

Abstract This chapter introduces the Beautiful Midden Field School Syllabus (BMFSS), an arts-led situated syllabus developed for students aged 16–18 years. Emerging from a short pilot project, the resulting syllabus integrates the extensive experience of an arts practitioner working with a range of media and diverse contexts, at the intersection of arts, technology, science and education. The BMFSS was developed to address limitations within STEAM education such as the need for a culturally situated approach, the inclusion of indigenous pedagogy, multiple world views and a systemic design-thinking perspective that integrates science.

The chapter is organised in three main sections. First, the chapter discusses the transition from STEM arts to STEAM education. This is followed by an introduction to the evolution of contemporary arts practice beyond traditional genres such as painting and sculpture, outlining the potential of, and insights from, emergent arts practices for STEAM. This serves to support the evolution between STEM arts and STEAM education through the inclusion of ecological design thinking, contemporary arts training and the integration of science themes. Finally, the chapter presents an overview of the core BMFSS syllabus, its rationale and methods. Central to the BMFSS is the permaCultural resilience (pCr) framework, (McKeown 2015) which integrates theory and practice to develop a systemic transdisciplinary approach to learning that is creative, situated and embeds eco-social justice at its core. The resulting syllabus, adaptive for age and ability, forms the basis of a model of praxis for a situated approach to STEAM education.

In conclusion, the chapter makes the case for an emergent, integrated STEAM syllabus, informed by an artist | educator's active engagement over 24 years with the complexity of a situated approach to teaching and learning.

Keywords Situated art practice · Post-studio art · Ecology

Artist | educator is used to indicate an integrated role that embeds education within arts projects, formally and informally as integral to the production of the resulting artworks.

A. McKeown (✉)
SMARTlab, University College Dublin, Dublin, Ireland
e-mail: anita.mckeown@ucd.ie

© Springer International Publishing AG, part of Springer Nature 2019

A. de la Garza, C. Travis (eds.), *The STEAM Revolution*,
https://doi.org/10.1007/978-3-319-89818-6_8

From STEM to STEAM

The advent of the STEM-art education movement raised a series of important questions for me as an artist I educator working at the intersection of arts, technology, science and education. What is the value and role of the arts within STEM-art education? Why is an arts-led approach different to STEM arts? How is the knowledge from a half-century of non-object-orientated art practices that are transdisciplinary and systemic integrated into this movement? How can situated and emergent arts practices that include science themes align to existing education standards? The BMFSS was an attempt to address the initially limited STEM-arts framework and to reposition the potential of art and design to contribute to a more integrated approach.

John Maeda, as president of the Rhode Island School of Design's (RISD), sought to place art and design at the centre of STEM and the STEM-arts movement through promoting a STEAM agenda (Maeda 2013; Lamont 2010). In January 2011 the National Science Foundation funded a 2-day RISD workshop entitled *Bridging STEM to STEAM: Developing New Frameworks for Art-Science-Design Pedagogy* (Rose 2011). This event brought together 60 experts working in science, IT, engineering, art and design, maths and education to progress an innovative educational agenda through an interrelational approach to art, design and STEM. The aim of the workshop was to develop transdisciplinary interactions that could lead to new teaching approaches to creative problem-solving (Rose and Smith 2011).

Transitioning from STEM to STEAM does not simply add art to STEM education or utilise the arts as a vehicle to teach STEM. Neither is it a matter of simple semantics; STEM, STEM arts and STEAM education have subtle nuances that should not be ignored: this transition promotes the creative risk-taking and exploratory processes inherent in art and design training and disciplines within the STEM fields. Within STEM arts, the arts can often be limited to traditional art forms that may have STEM topics as content, for example, designing a poster to communicate or 'about' STEM concerns or utilise an art form to teach STEM knowledge. By contrast, STEAM projects reflect contemporary interdisciplinary artistic concerns that are not bound by traditional media and utilise STEM skills and knowledge, such as electronic engineering, programming, or biological processes for their production while also including contextual studies, positioning the artworks within the larger canon of artistic practice. In other words, STEAM places value on the arts for their creative methodologies, ways of knowing the world and tangible modes of knowledge production. In addition, the arts can also disseminate STEM knowledge in a more accessible manner by 'making connections between diverse ideas and provok[ing] unexpected conversations' (Wellcome Trust 2017, para 3). This has particular relevance for the needs of a twenty-first-century post-disciplinary education system. Although distinct knowledge domains and specialisation will always have a place in education, there is a need to embrace exploration and experimentation. Within the century of the system (Gawande 2014), no field or discipline in isolation can adequately meet the needs and challenges of contemporary society. In STEAM education, learning occurs at the intersection of the five fields, transforming how we know and investigate the world. As a pedagogical innovation, the STEAM agenda offers an approach to teaching and learning 'that encourages and facilitates unorthodox methods and strategies' (Rose and Smith 2011, 8).

Beautiful Midden Field School Pilot

In 2016 I was invited to work with Amigos Bravos and their art and activism programme.[1] Amigos Bravos, a non-profit organisation based in Northern New Mexico, is a state-wide water conservation project located in a rural community in the high desert. Amigos Bravos partnered with the Beautiful Midden project (2012), established by a group of Taos artist/architect residents, to bring 'creative energy to issues and redefine public space in ways that the scientist and the environmentalist [communities] are unable' (2017).[2] In 2013, Scott Moore took the lead, evolving the Beautiful Midden project into a three-phase outreach programme of field trips, arts education and public presentation. In 2016, Scott left to continue his work in Vermont, and I began working in collaboration with Amigos Bravos to develop ideas for Beautiful Midden.

Scott's outreach programme underpinned the first phase of the field school. It was an exploratory introduction to the potential of the arts to engage students differently with the issues of water and waste. Two full-day educational events were undertaken in collaboration with LEAP, Amigos Bravos and Questa High School, 8th and 9th grades (13–15 years old). The events introduced the students to four artists: Mierle Laderman Ukeles, Chris Jordan, Vic Muniz and Swoon et al. who have worked with the concept of water and waste in different ways, namely, performance, sculpture, photography and film. This was followed by a field trip to local illegal dumpsites, the local recycling centre and hike down to the Rio Grande riverbed, a riparian ecosystem. The students were also introduced to the Bureau of Land Management (BLM) ranger, who presented the processes of land management in relation to the watershed. A second trip included an introduction to the acequia systems—history, management and benefits—and the full Rio Grande water shed by Miguel Sansistevan, an ethnobotanist, educator and mayordomo.[3] After lunch the group headed to a constructed lake and then onto the Rio Grande at the Red River fish hatchery to take water samples to compare and contrast. Each trip included student activity sheets provoking the students into making connections between the artists' work, the ideas behind the work and beginning to explore and consider the watershed and waste differently.

---

[1] See http://www.amigosbravos.com/art-activism

[2] www.beautifulmidden.org

[3] The elected overseer of a managed irrigation system known as an acequia, of which there are over 800 in New Mexico..

Local experts integrated knowledge from other fields into the fieldtrips, for example, the BLM ranger, the Amigos Bravos water sampling team, the recycling centre manager and the ethnobotanist. The combination of the artists' work, the arts activities, the field trips and the learning from the classroom initiated a discussion on the diverse forms of engagement with the subject matter. This process embedded the students' local context and personal experience as a foundation for their own learning and creative exploration. Students were also asked to devise questions for the experts they were introduced to as a means to explore their curiosity and introduce inquiry-based learning into the process. Activity sheets and verbal prompts encouraged the students to share their classroom knowledge of the water cycle and the riparian system. This helped contextualise their knowledge and apply it within a tangible experience and offered an opportunity to explore self-assessment. The students self-identified skill gaps, which in turn motivated student-generated learning (Jensen 1998; Wiggins 1997). In this instance, self-assessment became a valuable learning tool, and although not a key aspect of phase one, it has been integrated into the BMFSS. The process of engagement, team teaching and field trips led to the creation of the BMFSS as part of my work with on the evolution of the Beautiful Midden.

Towards an Arts-Led Syllabus

Before expanding on the core ideas and methods of the BMFSS, it is important to acknowledge a development in contemporary art practices and debates that is the foundation of the BMFSS approach to STEAM: the twentieth century saw artists increasingly motivated to engage and communicate through their work and connect with the lived experience of themselves and others. These desires necessitated a move out of the studio, travelling beyond the confines of the gallery to resituate work within the context of the every day. By the late 1960s, the post-modern concept of a 'post-studio' practice was commonplace. Artists' Placement Group, Robert Smithson and Suzanne Lacy are early examples of artists moving beyond the idealised object-orientated practices that had dominated the art world. This migration, referred to in the seminal text *Six Years: The Dematerialisation of Art* (Lippard 1973) and the continued shift away from the art object, is evidenced in the myriad of practices within Public or Live Art. Terms such as socially engaged practice (SEA), social practice (SP), context-responsive practice and relational (Bourriaud 2002) and dialogical aesthetics (Kester 2000) seek to further define the nuances of the extensive activities within an expanded field (Krauss 1979).

Not limited to aesthetic concerns, such projects also aim to address social issues, encourage relationships or develop conditions conducive to dialogue. They are often performative, use non-art materials and occur in non-art contexts, sometimes evolving over years. When such practices are place-based, they are considered situated practices, which can be understood as a family of artistic practices that transcend particular forms and genres. As such, these practices, without a fixed form, share the common root of artworks emerging from the situation or context in which the artist is situated. They are often but not always ephemeral and engage with social issues and take a critical stance that seeks to make change through action, reflection and active citizenship (McGonagle 2010). As 'art-in-the public interest' and (Kwon 2004) engagement with 'social issues, political activism and community collaborations' (2004: 60), they increasingly aim to perform what Jane Rendell calls 'critical functions' (Rendell 2008). As an activist-artistic practice incorporating dynamics of time, space, relationships and infrastructures including economic and political systems, a situated practice presents opportunities or junctures for new kinds of relationships with place. Artists working this way develop a diverse skill set over time through interdisciplinary practices and contexts, which benefits their delivery of STEAM education. Resulting artworks co-exist within the fields of ecology, politics and economics, and although firmly positioned as art, they consolidate a spectrum of 'cultural strategies that are experimental and experiential' (LADA 2015). As emergent practices, they may be unknown to those outside contemporary arts practices and therefore unrecognised as art. This can limit a non-arts practitioner to delivering STEM arts rather than fully realising the potential of STEAM education.

The permaCultural resilience (pCr) framework (McKeown 2015), a critical praxis—meaning a theoretical framework and practical toolkit—for situated art practices, is central to the BMFSSs. As a STEAM learning intervention, the BMFSS seeks to encourage a systemic understanding of a situation, disrupting the students' existing perceptions of art and context while enhancing their ability to intervene ethically. As such, the BMFSS draws on artistic practices that are activist and disruptive, that no longer seek to change the world or the whole of society. Instead, the aim is to initiate a ripple effect through interventions that are motivated by and target specific concerns. The BMFSS seeks to involve students in a learning process that engages with the 'temporal systems of organisations and public circulation' (Sholette and Thompson 2004:135) and catalyse an *art* | *activist* process as a disruptive influence that facilitates the emergence of new meaning through innovative communication channels impacting on attitudinal change (Holden 2015; Whitehead in Anderson and Prendergast 2012; Edelman et al. 1996). A process-driven activist practice extending beyond formal aesthetic concerns is a key aspect of the context in which the BMFSS sits. In this way, the BMFSS promotes a form of arts education that does not conform to conventional art forms used to teach STEM topics. An *art* | *activist* process situates students in systemic learning derived from their tangible experience, further implicating them through their personal creative response. This is an aspect often overlooked by those discussing or promoting STEM education through the arts.

The Beautiful Midden Field School Syllabus: An Introduction

The core of the BMFSS is rooted in permaculture,[4] an ecological design system that was adapted through doctoral research for situated art practices to create the permaCultural resilience (pCr) framework and practical toolkit. The pCr framework integrates art and ecological concepts into a systemic approach to making situated arts projects, which dynamically emerge from the context in which they evolve.

Among the skills that future employees will need are the following: the ability to utilise critical thinking, be technically competent, creative and possess the ability to adapt to a continuously evolving environment. These professionals will also be expected to 'operate outside the narrow limits of one discipline and […] be ethically grounded in solving the complex problems of the future' (Pierrakos et al. 2008:4). As new socio-economic and environmental conditions emerge, syllabi and curricula that address concrete challenges are urgent. The need to develop communication skills, inter- and cross-disciplinary knowledge transfer and collaboration across technical and non-technical boundaries is now an imperative.

Working within broader ecologies of practice more readily highlights, reflects on and challenges the complexity and dynamics of lived experience. The BMFSS acknowledges a systemic approach to change in which the intervention engages multiple actors and their knowledge for tactical approaches to develop longer-term strategies. As such, the BMFSS recognises the inadequacy of individual capacity to engage with little more than a singular aspect of a system. The process embedded within the BMFSS offers an opportunity to address scale through localised innovative action, creating initiatives and tackling concerns that are also global. The arts can support a tangible, physical exploration of abstract concepts, contributing to a more concrete personal understanding of issues that may lead to action. The development of a STEAM field school model of learning that sits alongside traditional modes of education offers an opportunity to pose new questions through creative problem-finding and project-based inquiry.

The BMFSS's ambition is to initiate processes that leverage the maximum resources available to contribute to framing and solving problems. Foregrounding the importance of understanding a problem from multiple perspectives acknowledges the perceptions and attitudes of those charged with solving the problem. Accurate problem framing and reconsideration of problems become an integral part of a problematic situation, with greater understanding heightening the potential of accessible, appropriate achievable solutions. The exploration and experimentation with new materials, ideas and concepts, without a fixed structure or defining theory inherent within art and design training, facilitate this process. Through visual,

[4] Originally employed to refer to a system of agricultural and social design principles simulating or directly employing the patterns and features observed in natural ecosystems, the term permaculture was expanded to stand also for 'permanent culture', since it was understood that social aspects were integral to a truly sustainable system. The use of a capital C in permaCultural resilience denotes the shift from a focus on growing food towards an intervention into culture to provide a systemic framework for use within public art practice.

structural, material and conceptual experimentation and research, artists and designers gain insights into possibilities through trial and error and embodied learning. The implementation of art and design processes to resolve artworks or fulfil briefs is often taken for granted and has proven to be significant method in problem-solving (Getzels and Cziksentmihayli 1965, 1976; Whitehead 2006). As a process of problem-solving, the BMFSS model challenges silo-specific approaches to education that create limited responses and continue to promote specialists or professional expertise as the only mode of problem-solving. The solution of complex systemic problems through looking at isolated parts of a system is ultimately unsustainable due to the growing scarcity of resources. By focussing locally, the BMFSS encourages the support of existing networks and administrative infrastructures and fully utilises the resources available.

The BMFSS has two modules: the Beautiful Midden Arts Lab 1 (BMAL1), a 20-week foundation programme, and BMAL2, an advanced level 12-week in situ programme. Students usually complete BMAL1 before undertaking BMAL2. BMAL1 focuses on skill building using the pCr framework, its methods and tools for developing, producing, presenting and disseminating artwork. The students also learn the foundations of situated and emergent contemporary art practices. The incorporation of situated art practices within the syllabus develops a creative, systemic transdisciplinary approach to learning, cross-pollinating and synthesising knowledge between disciplines.

In addition, bespoke artists' resources have been created around specific artists and concerns, namely, those working in sustainable practices, addressing water and waste (New Mexico indigenous and female artists). This introduces students to expanded arts practices and topics that many artists in the public realm are working with while increasing their knowledge of art history beyond traditional genres into a wide range of arts and related areas within technology and science. Being a culturally situated approach, this is also an important and locally meaningful way to engage with Common Core Standards and ensures that students become aware of the myriad of practices and practitioners that are present locally. Further, this enables students to see themselves and their cultures represented in the local, national and international cultural landscape. Recent scholarship affirms the importance of artistic expression as a means of 'developing a sense of personal and collective agency' (Ali et al. 1:2015) (Smith et al. 2012) and reflecting on personal identity, circle of influence and perceived worlds or realities (Gaskins 2010). Access to such representation also opens avenues and identifies through-routes for further study and professional practice.

Each week is thematically organised to focus on building students' knowledge across arts education, critical thinking, communication and discussion with practical activities and multimodal exploration of ideas around the issues raised. Blended and practice-based learning are used to teach the foundational aspects of situated arts practices and the key methods from the pCr toolkit. BMAL1 as the foundation focuses on key terms within expanded art practices, artists resources (water and waste), artist meet and greet (on and offline) and an introduction to the pCr's toolkit.

Arts activities including presentation, assignments, discussion and peer-2-peer assessment support the development of the students' skill set in preparation for BMAL2. Increased awareness of sustainability practices is integrated into BMAL1's syllabus to develop students' ability to engage with broader ecological issues in producing artworks as a foundation for BMAL2. BMAL2 focuses on a real-world project site developing their systemic approach to making situated arts projects. Working with a range of professional entities—e.g. local authority departments and public institutions—implements the knowledge they gained in BMAL1. This process embeds social and environmental justice in its methods enabling students to immerse themselves in a localised context, analyse the system and harness the various knowledge ecologies and perspectives for interpretation within any artworks or projects that emerge. Key to the emergent BMFSS was the push towards a truly interdisciplinary STEAM model beyond simply two or more disciplines in the classroom. This aims to encourage students to become actors working across a range of disciplinary frontiers to address global challenges through engaging with the earth's myriad of systems. Incorporating STEM knowledge into arts education offers a conduit for a transdisciplinary approach to learning. As a teaching and learning strategy, this process offers insights and processes from each discipline that creates new perspectives that can be incorporated into any resulting solutions. Further, this encourages a competency and confidence in the students to explore common connections and exploit differences, through their understanding and knowledge of the landscape's topography.

In addition, an artwork or the fulfilment of a design brief embodies a set of values that addresses materials, production processes, users and the greater ecosystem of society and its environment. Our systems and the objects we encounter daily are increasingly shown to be unfit for purpose, entrenched in inequitable ideologies and unsustainable production processes. There is a need for a re-imagination that acknowledges systemic interconnectedness and the consequences and impacts of many of our production processes and organisational infrastructures. This necessitates the creation of answers and a process of problem-solving that places humane responses that embeds social and environmental justice at the centre of teaching and learning. The BMFSS aspires to address both local and global scales through localised activities connected to global concerns. As an artistic praxis embodying 'thinking globally' and 'acting locally' (Dubos 1972 in Eblen and Eblen, 1994), student engagement becomes a localised tangible intervention into global issues that are deeply political.

BMAL1 begins to forefront artistic practice as a means to pose pertinent questions, bring to light new perspectives on scientific and environmental issues and derive new approaches to engaging with these concerns. Within the foundational module, students are introduced to key movements and examples of emergent art practices and an introduction to the debates around the concept of 'site'. This expands their understanding of contemporary arts practices and the concerns they will have to address within their projects. These aspects are neglected within current arts education, which focuses on traditional skills and genres.

BMAL2 enables students to build on the foundational module through the practical experiences of working on a real-world project site, practicing the skills they learnt in module 1 and learning additional tools that are part of the pCr toolkit; zoning and the Vital Signs Evaluation Matrix. The zoning and Vital Signs Matrix, in conjunction with the OBREDIM mnemonic, builds a complex mapping of the cultural, economic, socio-political and environmental dynamics.

The BMFSS Toolkit: The OBREDIM Mnemonic

The integration of arts with an ecological approach to a context encompasses disciplines as diverse as environmental sciences, economics, sociology and biological and environmental engineering. The pCr's design principles facilitate the close observation necessary for drawing attention to a broad range of inputs, contributing to a unique assessment of the location. BMAL1 and BMAL2 both use a systems design method, the pCr toolkit, which focuses on engaging students with a real-word situation or context, through a deep audit. In BMAL1 an introduction to exploring each aspect of the OBREDIM mnemonic (see Table 8.1) is facilitated through breaking down project examples. This supports students to develop the pCr audit skills through a critical engagement and analysis of the projects' context and production processes. In BMAL2 this is the initial process by which the students begin to develop their own projects in response to the specifics of the real-world site they are working on.

The OBREDIM mnemonic facilitates the gathering of objective and subjective data by providing two field logs, as well as facilitating the discovery of anecdotal and historical, factual information, which are then playfully and experimentally remixed within the students' projects. The mnemonic also provides a guiding framework for an itinerant idiosyncratic process that reveals new perspectives and multiple understandings and encourages the students to reconsider the context and any 'problems' or issues that are encountered within the project site.

Finding the problem and developing locally grown solutions by understanding and imagining things differently are integrated into the BMFSS' teaching and learning methods. As a facilitated arts-led learning process, the students' imagination is harnessed within a facilitated process of critical thinking to provoke a reconsideration of a known context. This also lends itself to a method of problem finding, a practical articulation of the steps in problem processing—shaping, defining and solving. This requires the application of creativity and vision from students, what Donald Winnicott has called 'creative illusion' (1967), an imaginative process that precedes change through the ability to imagine things differently. This has also been called associational thinking and is considered, along with questioning, observing, networking and experimenting (Dyer et al. 2009), as one

Table 8.1 OBREDIM Log 1. (Adapted McKeown 2008–2015) forming the pCr audit (McKeown 2015)

OBREDIM AUDIT	Activity Date
Audit phase	Details reference: What to look for and record
O—observation **Phase 1**	Survey all local organisms, e.g. organisations, stakeholders, businesses, arts and cultural orgs, community groups, charities, people, animals, vegetation, sociocultural landscape, history, news/media and politics. Try to create as full a picture of the residency ecosystem
Ways and things to observe: patterns of growth; distribution, town layout, business layout, etc.; traffic flow, people motion, dead spots, flow of information, traffic people, the dynamics; social, cultural and physical. Is there an impact? Does it last? Where does it start and stop?	
Natural system aspects: weather, sun, watersheds, air, flora fauna animals, migration routes or diversions of water, desertification, forest	
History—what has changed and why is there a pattern, does this have impact on the future? Communities, connections and relationships. Distance/proximity, interspecies, intergenerational. What is successful? What have adapted and are there any common traits?	
Are there any recognisable patterns and numerical patterns? Are their functions of these patterns? Look at textures/shapes—draw them, photograph them, and record audio and video. Use the senses: what can you see, hear, taste, smell and touch?	
B—boundaries	The edges/limits of the ecosystem; the location's geo boundaries, organisational boundaries. People's responsibilities, shared values, cross-over of aims, power dynamics. Limits to growth expansion. Laws, regulations and policies. Where do things stop and start? Are there diversity, tensions and encounters? Is there a difference between the edges and centre? Zoning analysis: This can highlight responsibilities, existing partnerships and fees for effort
R—resources	Physical and non-physical resources; time, money, services, skills and knowledge, existing networks and partnerships, groups, what already exists and how it works (or doesn't) Sample questions: venues, what's in there, what does it do, how does it function and who sponsors events. Groups: who's doing what, when and with who
E—evaluation **Phase 2**	Begin to map a web of relations—using the info from phase 1 and the zoning analysis. Evaluate what exists and where the gaps are—How does info flow, notice relationships and communication. Include a SWOT/ SMARTER analysis
D—design	Design on paper. Becomes a map for the implementation stage or if there's an existing project in mind redesign in light of information gathered in phase 1 and evaluation stage.
I—implementation **Phase 3**	Implementing design including logistics, e.g. timelines, production milestones, communication, fundraising, skills needed
M—maintenance	Maintaining the project and any maintenance needs or opportunities to evolve the project, handing over passwords, admin details, resource directory—anything needed to move the project forward or maintain its existence and evolve it

of the key skills of positive disruptors (Christensen 2014). Dyer et al.'s extensive research[5] found that creativity is a function of both the mind and behaviour concluding 'that if we can change behaviours we can change our creative impact' (Frogget et al. 2011:3).

The OBREDIM mnemonic supports a review of the broader ecosystem and the context in which the students operate. Application of the mnemonic provides a useful tactic to facilitate, gathering the ecosystem's local knowledge, an asset often undervalued. In combination with an artistic skill set, this knowledge is actualised within any project's ecology. Using the OBREDIM method ensures the skills for assessing a localised context, including local policy issues and regulations, which are shared with students. As an empirical method, the skills are also beneficial for undertaking scientific inquiries, e.g. observation, data collection and analysis.

By encouraging the students to develop arts interventions, an opportunity 'to break an established order and to allow the possibility of new connections and ideas' (Frogget et al. 2011:63) is integrated into the field school's learning methods. The BMFSS interrupts the students' understanding of a situation or context and serves to enable new perceptions and opportunities for action by reimagining the context through an eco-social cultural positive disruption. This empowers students to enact and express their concerns around socio-political and ecological issues within a public platform as a means of agency from which many are often excluded.

The BMFSS Toolkit: Zoning

Permaculture's zoning principle highlights relationships within the project and is used for two aspects: mapping the project's constituents and components and a simple lifecycle analysis of the project's inputs, processes and outputs. Mapping the relationships between various participants, organisations and their relationships with the physical environment aids organising aspects of a project to identify potential partners. This builds the projects' micro-ecology with a context mapping process, extending out from the central point of the project (Fig. 8.1).

Mapping the potential micro-ecology for integration into a project enables the students to visualise the interrelationships, reflecting on the project context's distinct nature and encouraging small-scale targeted actions. The mapping is reviewed at the evaluation stage of the OBREDIM framework, as part of conceptual planning and design of the project.

---

[5] Dyer's research gathered data from 500 business innovators and over 5000 executives in 75 countries.

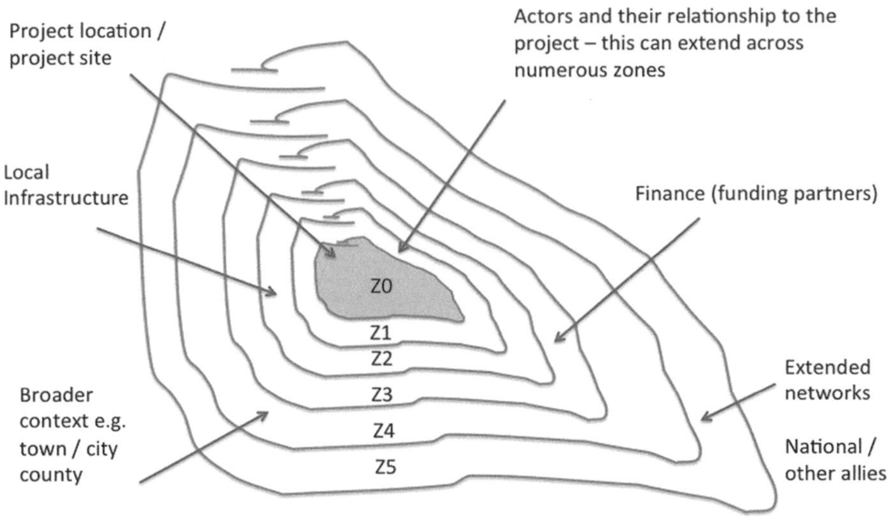

Fig. 8.1 Generic zoning tool example showing simple project mapping

The BMFSS Toolkit: Life Cycle Analysis

The zoning tool also enables a simple lifecycle analysis by drawing attention to and identifying issues of scale; further away from the focal point of the project, the zoning represents a larger area, and consequently more energy/effort inputs are required. For example, if the mapping shows the majority of the activity is in zone 5 or materials are coming from zone 5, then perhaps there needs to be more consideration around embodied energy. This may appear in transport or the ethics of sourcing and production, which may incur additional energy and waste into the process. Once the project's inputs, processes and outputs are mapped, the students' awareness of the project's ecosystem increases and develops their systemic thinking.

The zones also reflect proximity or intensity in relation to the project including relationships and dynamics of accountability and responsibility. This is an important aspect of project development and management. Students gain the ability to recognise over extension/ambition which can undermine their projects. Zonal mapping identifies areas of concern and guards against the dissipation of effort. Students are empowered to make decisions during the evaluation and project design stage, for example, whether to continue developing certain partnerships or use certain materials.

Students can also decide to use the outlying assets as a new project node from which to initiate the mnemonic. These can form new 'ripple' effects and can carry information or the impact of the project into new arenas or legacy projects. The projects can also be considered as cellular in nature, reproducing once capacity has been reached rather than an endless expansion. The cell or project 'splits' to form new related projects that can mutually support each other, increasing potential of production through supporting self-reliance and increased interdependence.

The BMFSS Toolkit: Vital Signs Evaluation Matrix

The final tool that comprises the BMAL2 module is the Vital Signs Evaluation Matrix, an evaluation and planning tool that also embeds social and environmental justice into the students' projects. There are two aspects to the matrix, permaculture's triad of capital Earth care, people care and fair share (see Fig. 8.2) and the vital signs of a pCr project (Table 8.2).

As a project-planning tool used within the design phase of the project's development, the VSM enables students to consider aspects of their project across each criteria in the matrix and how they might bring these values into their projects. These can be aspirational, identifying values that may be important to work towards but that are not feasible at this stage. The VSM can also be used to assess completed projects developed in dynamic 'non-lab' contexts where causality is impossible to prove. The VSM can be used to assess if values that are set at the beginning of a project have been met as well as to identify the tangible outcomes of a project.

Through the BMFSS students begin to develop a complex and systemic understanding of the context in which they are working and its broader global position. The additional experts and practitioners that contribute to the syllabus delivery support this development. Students gain the foundation to develop comprehensive STEAM projects utilising the knowledge from their evaluation of the context, which evolves using the pCr tools.

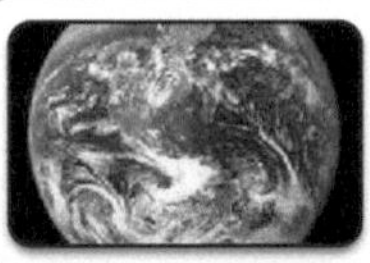

Earth Care: Considers the diversity within the project's ecology and engages with consumption assessing materials and services. Also has an additional awareness for bio-centricity and recognising the need to move towards eco-centric practices.

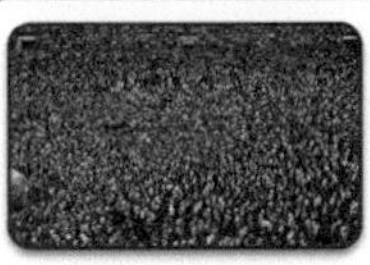

People Care: Considers the accessibility, inclusivity, reciprocity, collaboration, participation within the project and processes. Also acknowledges the value and diversity of multiple knowledge cultures.

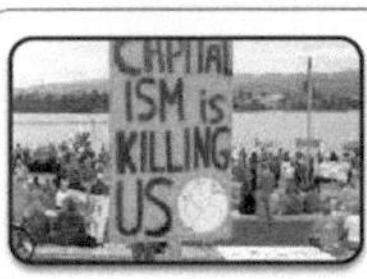

Fair Share: Considers the ethical working practices of the projects and their ability to contribute towards a sustainable ecology including sharing surpluses eco-social responsibility and accountability. Can also include a full life cycle analysis of businesses if appropriate / possible.

Fig. 8.2 Permaculture's triad of capital

Table 8.2 The Vital Signs Evaluation Matrix showing generic values

	Permaculture triad		
Vital signs	Earth care	People care	Fair share
Building micro-ecologies	Sustainable practices and materials diversity	Multiple entry points. Accessible	Diverse outputs and equitable exchanges
Strategic intervention Tactics	Interventional/process-based agile and adaptive	New policies or operational procedures	Bio/ecocentric awareness Sustainable resource management
Reseeding local Knowledge	Sustainable and purposeful	Knowledge equity, new networks and self-organisation	Agency. P2P, open source
Resituating arts and culture	Creating projects and developing legacies and proposals with eco-social commitments	Collaborative, participatory, imaginative and innovative	Transferable/sharable systemic and situated approach

BMFSS Assessment

The syllabus evolved within a US context, which does not have a national curriculum. In 2009 a set of national standards, the Common Core Standards, set out benchmarks for high school students.[6] The syllabus integrates New Mexico's Common Core Standards for Visual Arts (NMCSVS); however, the Common Core Standards do not cover science or social science. To address this content gap, the Next Generation Science Standards (NGSS) were released in 2012 and have been adopted by many states.

The alignment to the Next Generation Science Standards within the BMFSS encourages the development of key STEM skills through related subject areas, for instance, ecology including water and waste science, phytoremediation, recycled art and e-waste. The syllabus has been structured to foster student exploration of technology and engineering within their projects when appropriate. This also includes the integration of physical computing such as sensors, Arduino and, when available, 3D printing and laser cutting. Through the incorporation of the lifecycle analysis within the syllabus, students learn to assess their project life cycle, and the impact of any technology included and makes context-responsive decisions.

Embedding a practice-based approach to learning aims to develop a student's skill set and critical thinking by utilising the tools and techniques of pCr framework. This includes research into the work of other artists in the field, cross-disciplinary collaboration and individual and group project development. Students' practical activities and their self- and peer-to-peer assessment assignments evidence their

[6] Forty-two states adopted the Common Core Standards, which initially only covered English and maths. Since adoption three states have decided to repeal or reform them. In an attempt to create a national standard, the adoption of Common Core has been tied into schools' funding streams. The current administration plans to abolish the standards but as yet nothing has been confirmed.

learning against the New Mexico Common Core Standards and Benchmarks. A range of methods and skills important for other fields are also developed through the syllabus' assignments including:

- Presentation skills
- Critical and peer assessment
- Project design/implementation/management skills including health and safety issues and risk assessments
- Developing a network of contacts
- Identifying and assessing opportunities including applying for grants, residencies and college programmes

Self-assessment has been included with peer-to-peer (p2p) assessment as a means to promote and motivate meaningful learning and a deeper understanding of the assessment process (Bruce 2001). Both forms of assessment are common within arts education in combination with reflective practice, enabling students to structure their own learning as well as internalise the criteria for success. Students are supported to make new connections and enhance learning in a personal and meaningful way (McMillan and Hearn 2008) by encouraging mastery over performance (Dweck 1986). Mastery, as a key motivating factor for learning, does not rely on teachers to coordinate learning tasks and facilitate students to generate their own solutions. Although the BMFSS is aligned to the performance goals of NM's Common Core Standards, mastery is also an important aspect of arts education assessment. The syllabus' methods encourage mastery by immersing the students in the subject matter from a range of perspectives. The students' learning is integrated into the production of an art outcome that embodies their thinking and learning through a tangible process. By supporting students to gain these skills, the BMFSS was developed to support students to become more confident, competent and motivated to challenge themselves to complete difficult tasks (Rolheiser and Ross 2000, 2001).

Conclusion

As a flexible and adaptive model for learning, the BMFSS's strength comes from the integration of multiple knowledge cultures and adaptive content, both practical and theoretical, to engage students in a dynamic situated process. This aids the development of systemic critical thinking and opportunities to extend such thinking into other disciplines and integrate field-specific knowledge into a broader learning ecology. This approach seeks to encourage students to be creative and experimental, posing new questions or interrogating old questions and concerns in new ways.

Forging new pathways towards new thought-paradigms that are grounded in eco-social justice offers a unique opportunity to develop resourceful solutions that are relevant. The artists' resources that accompany the practical exploration serve to introduce students to a wide-ranging experience of post-studio and non-object-orientated practices and ways of engaging with different themes. Incorporating local practitioners enables the students to 'see themselves' represented culturally and within an evolution of art-making practices and begin to situate themselves within a professional context.

The exploration of a thematic, experimental learning model offers education a context-responsive framework that can be aligned to existing standards yet is not bounded or inhibited by such standards. Alignment with the Next Generation Science Standards and New Mexico's Common Core Standards for Visual Arts within the BMFSS offers an opportunity to expand existing standards to incorporate new technologies, contemporary art practices and pedagogical developments in an agile way. The inclusion of scientific content, emergent art practices and design thinking is currently absent from second-level arts education. The integration of both these standards facilitates uptake within the public school system and access to the necessary funding to enable comprehensive trialling.

Initially developed for students aged 16–18 years, the BMFSS can be delivered formally within the school context or to augment formal learning through informal settings such as summer schools, youth groups and arts organisations. The BMFSS syllabus can also be adapted for more advanced students, for example, in advanced programmes, undergraduate and postgraduate as well as for younger students with the various learning resources taking into consideration age, ability, themes and location. The method can be aligned to other learning standards to ensure the students learning meets the need of any education department's curriculum. The pCr framework underpinning the syllabus design promotes social justice, fostered through an adaptive syllabus and inclusive design principles to address venue access, language, delivery and when necessary physical and cognitive ability.

Practitioners working at the interface of arts, STEM and education all agree that there is a critical need for empirical knowledge (Seifter in Robelan 2011) of what the arts bring to the STEM subjects even though there are many first-hand and anecdotal accounts of the benefits. A full trial of the model has been postponed until September 2017 due to the impact of the US administration on the arts and education funding. There are plans to adapt the model for other contexts and content to evaluate its transferability. Discussions with other practitioners are also underway to encourage their explorations.

As more students are trained in innovative multidisciplinary art and design thinking, it is reasonable to assume based on existing models and human evolution to date that creative solutions could contribute to an equitable and sustainable future. By exposing students to multiple art and design examples with a practical exploration, it is hoped that the BMFSS scaffolding and its core methods encourages students to experiment more within their thinking and application of knowledge. It is too early to tell as more rigorous research is needed, but the syllabus aspires to inspire new questions that lead to fresh solutions and innovative approaches and ways of thinking that are relevant for the twenty-first century.

References

Ali A, Sharp S, Meged S (2015) An empowerment-based experiential learning model: The PhotoCLUB program for children and adolescents. Journal of Art for Life, [Accpeted/In Press]. Available online https://nyuscholars.nyu.edu/en/publications/an-empowerment-based-experiential-learning-model-the-photoclub-pr. Accessed 15 Apr 2016

Amigos Bravos (2012) Beautiful Midden. Available online http://www.amigosbravos.com/art-activism. Accessed 18 Apr 2017

Anderson A, Prendergast MS (2012) The conversation: in search of the new normal. In Long Conversation: Frances Whitehead C5. Available at: http://www.findtheconversation.com/episode-25-frances-whitehead/. Accessed on May 15 2016

Bourriaud N (2002) Relational aesthetics. Presses du reel, Paris

Bruce LB (2001) Student self-assessment: making standards come alive classroom leadership. Sept 2001, vol 5(1). Available online. http://www.ascd.org/publications/classroom_leadership/sept2001/Student_Self-Assessment.aspx. Accessed 15 May 2015

Christensen CM (2014) Disruptive innovation. In: Soegaard M, Dam RF (eds) The encyclopedia of human-computer interaction, 2nd edn. The Interaction Design Foundation, Aarhus Available at: https://www.interactiondesign.org/encyclopedia/disruptive_innovation.html. Accessed on Nov 15 2014

Dweck CS (1986) Motivational processes affecting learning. Am Psychol 41(10):1040–1048

Dyer JRG, Johansson A, Helbing D, Couzin ID, Krause J (2009) Leadership, consensus decision making and collective behaviour in humans. Philos Trans R Soc B 364(1518):781–789

Eblen RA, Eblen WR (eds) (1994) The encyclopedia of the environment. Houghton Mifflin, Boston/New York

Edelman M, Weakley A, Candy L, Fell M, Knott R, Pauletto S (1996) From art to politics: How artistic creations shape political conceptions. University of Chicago Press, Chicago

Frogget L, Little R, Roy A, Whittaker L (2011) New model visual arts organisations & social engagement. Psychological Research Unit: University of Central Lancashire, Preston

Gaskins N (2010) Art & the avatar: ambiguity of identity in Virtual 3D Worlds, art21 April 21 2010 Available at http://magazine.art21.org/2010/04/21/art-the-avatar-ambiguity-of-identity-in-virtual-3d-worlds/. Accessed on May 15 2016

Gawande A (2014) The century of the system, Lecture 2, The Reith Lectures, The future of medicine. Available at http://www.bbc.co.uk/programmes/b04sv1s5. Accessed: 6 Dec 2014

Getzels JW, Csiksentmihayli M (1965) Creative thinking in art students: the process of discovery, Co-operative research report no. S-080, Chicago, University of Chicago

Getzels JW, Cziksentmihayli M (1976) The creative vision: a longitudinal study of problem finding. Wiley, New York

Holden J (2015) The ecology of culture: culture value project. Arts and Humanities Research Council, Swindon

Jensen E (1998) Introduction to brain compatible learning. The Brain Store, San Diego

Kester G (2000) 'Dialogical aesthetics: a critical framework for littoral art aesthetics' In Variant Issue 9 Available at http://www.variant.org.uk/9texts/KesterSupplement.html. Accessed on May 15 2010

Krauss R (1979) Sculpture in an expanded field. In October, vol 8, no. Spring, pp 30–44

Kwon M (2004) One place after another. MIT Press, Cambridge, MA

LADA (2015) What is live art? Available at: http://www.thisisliveart.co.uk/about/what-is-live-art/. Accessed on Aug 17 2010

Lamont T (2010) John Maeda: innovation is born when art meets science The Guardian 13th Nov 2010 Available online https://www.theguardian.com/technology/2010/nov/14/my-bright-idea-john-maeda. Accessed 11 Apr 2015

Lippard L (1973) Six years: the dematerialization of the art object from 1966 to 1972. University of California Press, Berkeley

Maeda J (2013) Artists and scientists: more alike than different Scientific America July 11, 2013 Available https://blogs.scientificamerican.com/guest-blog/artists-and-scientists-more-alike-than-different/. Accessed: 11 Apr 2015

McGonagle D (2010) Passive to active citizenship: a role for the arts conference paper – Bologna in context conference Dublin 24th Oct 2010. The Honorable Society of King's Inns, Dublin

McKeown A (2015) Cultivating permaCultural resilience: towards a creative placemaking critical praxis. PhD thesis, National College of Art and Design

McMillan JH, Hearn J (2008) Educational horizons, 87(1):40–49, Fall. Available online https://eric.ed.gov/?id=EJ815370. Accessed 15 Apr 2016

Pierrakos O, Kander R, Pappas E, Prins R (2008, January) An innovative engineering curriculum at James Madison University: Transcending disciplinary boundaries through innovative problem based learning practices. In: ASME 2008 International Mechanical Engineering Congress and Exposition. American Society of Mechanical Engineers, pp 189–195

Rendell J (2008) Jane Rendell 'Critical Spatial Practice' art incorporated, Kunstmuseet Koge Skitsesamling, Denmark. Available at http://www.janerendell.co.uk/wp-content/uploads/2009/06/critical-spatial-practice.pdf. Accessed 20 Jan 2014

Robelan EW (2011) STEAM: experts make case for adding arts to STEM http://www.edweek.org/ew/articles/2011/12/01/13steam_ep.h31.html

Rolheiser C, Ross JA (2000) Student self-evaluation—what do we know? Orbit 30(4):33–36

Rolheiser C, Ross JA (2001) Student self-evaluation: what research says and what practice shows. By Available online www.cdl.org/resource-library/articles/self-eval.php. Accessed 11 Apr 2015

Rose C (2011) Bridging Stem-to-steam. [ONLINE]. Available at: http://cjvrose.com/wp-content/uploads/2012/03/stem-to-steam-report.pdf. Accessed 12 Apr 2017

Rose C, Smith, B (2011) Workshop: bridging STEM to STEAM: developing new frameworks for art/science pedagogy Rhode Island School of Design, Jan 2011

Sholette G, Thompson N (eds) (2004) The interventionists: users' manual for the creative disruption of everyday life. The MIT Press, Cambridge, MA

Smith L, Bratini L, Appio LM (2012) 'Everybody's teaching and everybody's learning': photovoice and youth counseling. J Couns Dev 90(1):3–12

Wellcome Trust. The arts. Available at https://wellcome.ac.uk/what-we-do/our-work/arts. Accessed 10 Apr 2017

Whitehead F (2006) What do artists know? Available at http://embeddedartistproject.com/whatdo-artistsknow.html. Accessed 17 Apr 2013

Wiggins G (1997) Feedback: how learning occurs. Paper presented at the America Association of Higher Education Conference on Assessment and Quality, Miami Beach, FL

Part III
Structuring STEAM: Facilitating and Supporting STEAM Collaborations

Chapter 9
Retreating for Interdisciplinarity: The Case of the Health Research Futures Lab, Limerick

Annmarie Ryan, Niamh NicGhabhann, Muireann McMahon, and Jazmin Chodi

Abstract The need to develop interdisciplinary approaches to complex problems and societal challenges is evident across current and emerging policy within the EU. However, while interdisciplinarity is an important goal, the full potential of such an approach has yet to be realised. This chapter documents one approach to address the challenges of interdisciplinary working between researchers across AHSS and STEM, by charting one response to the dearth of work looking explicitly at the process of interdisciplinary working. It describes the design and implementation of the Health Research Futures Lab held in Limerick over 4 days in March/April 2016, which was based on a model of interdisciplinary research developed at the University of Limerick by an interdisciplinary group from the faculties of science, technology, engineering and mathematics (STEM), the Kemmy Business School and the Irish World Academy of Music and Dance (AHSS). In this chapter the design principles of the HRFL will be explained and explored in terms of their contribution to knowledge in the field of interdisciplinary research and the support they can bring. Key dimensions will be elaborated upon, including the use of design tools to enable researchers to identify and explore research ideas together; the retreat dimension, which meant that researchers worked over 2×48 h time frames researching, eating and socialising together as a means to develop community; and finally, the use of experiential exercises throughout the early stages of the workshop, which aimed to enable trust and build social ties.

A. Ryan (✉)
Kemmy Business School, University of Limerick, Limerick, Republic of Ireland
e-mail: annmarie.ryan@ul.ie

N. NicGhabhann
Faculty of Arts, Humanities and Social Sciences, University of Limerick, Limerick, Republic of Ireland

M. McMahon
School of Design, University of Limerick, Limerick, Republic of Ireland

J. Chodi
University of Limerick, Limerick, Republic of Ireland

© Springer International Publishing AG, part of Springer Nature 2019
A. de la Garza, C. Travis (eds.), *The STEAM Revolution*,
https://doi.org/10.1007/978-3-319-89818-6_9

Keywords Interdisciplinary practice · Academic retreats · Design approach · IDR communities

Introduction

The need to develop interdisciplinary approaches to complex problems and societal challenges is evident across current and emerging policy within the EU. These complex problems demand the dynamic, continuous and ongoing problem-solving process that crossing disciplinary boundaries can offer. However, as Catherine Lyall et al. (2015 pg.1) suggest 'interdisciplinary research does not occur automatically [...] It is not a simple case of aggregating several disciplines into one research project. Extra effort is required to achieve the promise of synergy and to form a genuinely cohesive team that combines expertise from several specialisms'. This is the focus of this chapter and indeed of the Health Research Futures Lab that it describes.

The Health Research Futures Lab (hereafter the HRFL) was designed to reflect and respond to the increasing emphasis on discourses of creativity and AHSS/ STEM interdisciplinary collaboration in addressing a wide range of societal issues in the contemporary world. The principles underpinning the design of the HRFL can be found at the confluence of three key insights into contemporary academic work: (1) the role of structured interventions to support academic practice (Moore 2003), (2) the value of experiential exercises to enable reflective practice and effective research team formation (see O'Malley and Ryan 2006) and (3) the role of design tools to enable new groups to cocreate solutions. The view of the design team was that the HRFL's workshop-based approach, which focused on both developing solutions to identified issues and on developing models for facilitating successful creative collaborations would enable interdisciplinary researchers to pursue interdisciplinarity in their working practices.

The design and organisation of the HRFL, held in Limerick over 4 days in March/ April 2016, was based on a model of interdisciplinary research developed at the University of Limerick by an interdisciplinary group from the faculties of Science and Engineering (STEM), the Kemmy Business School and the Irish World Academy of Music and Dance (AHSS). Findings from prior interdisciplinary labs, run by the team, were that barriers around trust, language, and existing preconceptions hinder greater collaboration. We assert that without an exploration of these barriers, and the development of models to facilitate this collaboration, the current low rate of AHSS/ STEM collaboration will be difficult to change. The HRFL aimed to offer a neutral space where knowledge and practices from each discipline were considered equal. Rather than offering one discipline in the service of another, the HRFL aimed to facilitate the integration of techniques, tools and perspectives from two or more disciplines towards identifying and addressing complex problems beyond the scope of any one of these disciplines. The main aims of the HRFL were:

1. To build capacity amongst Irish researchers to work in interdisciplinary research teams by facilitating an environment that supported interdisciplinary research, developing a structured approach to ideation, cocreation and co-researching
2. To bring a series of new research projects to pitch stage

The HRFL acted as a live experiential learning environment, where full time researchers were enabled to develop their capacity to work in interdisciplinary teams in a supportive environment. Researchers were facilitated through a design-led process, supported by the consortium team as well as research support staff. Participants began the workshop as individual researchers, and finished as members of interdisciplinary teams, having presented a proposal for a future research project to an expert panel. The approach taken made the workshop highly interactive and a productive use of researchers' time.

In this chapter the design principles of the HRFL will be explained and explored in terms of their contribution to knowledge in the field of interdisciplinary research and the support they can bring. Key dimensions will be elaborated upon, including the use of design tools to enable researchers to identify and explore research ideas together; the retreat dimension, which meant that researchers worked over 2×48 h time frames researching, eating and socialising together as a means to develop community; and finally, the use of experiential exercises throughout the early stages of the workshop, which aimed to enable trust and build social ties. These aspects, we believe, were vital in the overall success of this interdisciplinary workshop, and point towards the importance of not overlooking process when supporting IDR. In the following section, the main body of the chapter, we outline the process of designing and implementing the HRFL, including the recruitment of participants, role of design tools, experiential exercises, research support staff, participant reflection and the retreat nature of the HRFL. This is followed by an overview of the key outcomes. The chapter concludes with a reflection on key insights from the HRFL for the practices and process of interdisciplinary research.

The Health Research Futures Lab (HRFL)

A main aim of the HRFL was to facilitate an environment that supported interdisciplinary research, developing a structured approach to ideation, cocreation and co-researching, and to bring a series of research projects to pitch stage. This was achieved through a structured programme for the cocreation of new research project proposals with an iterative approach that encourages continuous change and refinement throughout. The success of the HRFL in achieving this aim is evident in the six final projects developed to pitch stage and presented to the panel of expert judges on the final day of the 4-day HRFL. The positive feedback from this panel of expert judges reflected the strength of the proposals. Seed funding was awarded to the teams with the most promising proposals and those that offered the best opportunity for interdisciplinary team working. In the design of this workshop the following elements were included to overcome these issues, and give time and space to researchers to form new research teams:

- Compulsory 4-day residential, spread out over two sessions.
- Teams were mixed from the outset in terms of disciplines and levels of experience.
- The group were introduced to various interventions to support their work and help develop social capital amongst the group as a whole. These included movement and trust exercises.

It should be noted that by the end of these 4 days, we had five research teams, each submitting a 2000 word new interdisciplinary research proposal. Each team made a final presentation of their work and three were shortlisted for seed funding prizes to support the follow-on of the work developed. The topics of the final proposals included informal care delivery, community maternal health service delivery and coping with lived experiences of chronic pain. How was this achieved? We address this question in the following sections.

Overview of the Approach to the *HRFL*

The team behind the HRFL came from a range of disciplinary backgrounds, including business, performing arts, fine arts, participatory arts, medical humanities and design. The process of designing and organising the HRFL became a microcosm of the challenges of interdisciplinary work. For example, because the organising team work across schools within the University of Limerick as well as another institution, this meant that each had different teaching schedules and constraints on available time. We thus had to meet sometimes as a group, and at other times only the Principal Investigator of the project travelled around to each member to work on specific aspects of the HRFL relating to their expertise. The team also made use of a shared work platform to ensure that all materials were accessible to all group members at all times and where documents could be easily co-written. Tasks were assigned to different members based on their experience, and each member of the group was directly involved in writing the call for participants, setting the evaluation criteria, publicising the event within their network, evaluating and selecting final participants (Section "Recruitment and selection"). For the sake of clarity and consistency in communication, the PI was the single point of contact for all would-be participants.

In designing the HRFL, various models of academic practice were applied. Firstly, following our previous experiences in interdisciplinary working the organising group felt that bringing researchers away from their day-to-day work spaces and routines would enable them to experiment and be open to new ideas and ways of working. It was important then that the location of the HRFL was a neutral one, and not on the 'physical grounds' of any one discipline. In addition, we felt that the participants should be facilitated through the experience, and the boundaries managed. For example, selection was contingent on being able to fully commit to the full 4 days of the HRFL, including overnight stays (there were some exceptions for 1 or 2 h, but this had to be agreed upon by the team). In being residential over 4 days the HRFL was following the model of academic retreats as espoused by Professor Sarah Moore (2003) at the University of Limerick. In her work, Professor Moore advocates the support of academic practice and acknowledges the challenges academics face in finding the time and space for writing. As a response she has developed, along with Rowena Murray, a model of academic retreats, which offer a structured and supportive environment for researchers. This is achieved by building an atmosphere of trust and safety, supporting the formation of a community and enabling within group learning (ibid). We extended this model for the purposes of developing new research projects and codesigning these in new interdisciplinary teams.

Recruitment and Selection

Participation in the HRFL was competitive, with each applicant making a case for why they should be able to participate. A key criterion for participation was an openness to engage with other disciplines. Each researcher was asked to propose potential research projects requiring an interdisciplinary lens, including current/live calls from research agencies. From these the organising group formed teams.

There were 44 applications from across AHSS, STEM and including a number of (independent) Fine and Performing artists. We were very pleased with the cross section of applications and final participants. We were not expecting as many visual/performing and art therapists, and working with individuals that were not traditional academic researchers meant that we had to consider our language and framing of the workshop to suit this particular cohort.

Each member of the organising group engaged in the shortlisting process, using an agreed grading scheme, which included two levels. One, a general merit score, which involved grading the applicants (1–3) on each of the following: previous interdisciplinary experience; previous experience of (interdisciplinary) team working; previous research in area of Health (futures); previous funding application experience; level of interest expressed/ambition to participate in future funding calls. Two, a grading scheme that allowed the design team to consider the overall make up of the final group. This was achieved by considering various criteria devised from a review of the literature on interdisciplinary research team formation and consisted of evidence:

- Of ability to 'speak multiple languages'
- Of empathy towards approaches other than just one discipline
- Of novelty of their approach/methods used (novel in the context of their own discipline)
- That they could 'fit' or find a group working in a similar area/issue

From the 44 applications, 26 were offered places, 23 accepted places, 23 started the first day of the workshop and on the first day one participant left. The following are the list of participants' disciplines: Food Science; Public Health; Visual/participatory Arts (2); Art Therapy (3); Physiotherapy; Medical humanities; Health/visual art; Humanitarian aid/visual arts; ICT; Physical Health; Design; Dance (2); Visual art/patient advocate; Psychology/participatory Art (practice); Management/Business; Psychology (academic); Creative Arts Facilitator.

Overview of Design Tools

The design of the HRFL was based on several models developed at the University of Limerick through projects like the Intelligence Unit [IU] with the School of Architecture UL; Health Futures HRFL and the 24-h Sustainability Challenge. Over the past 3 years these projects have explored different interdisciplinary configurations and focused on addressing complex problems at the regional level, in the areas of health, culture and the environment. What has resulted is a collaborative model that includes the building of a rich network of stakeholders, comprising public and private actors, to act as real-time support and input into the process. It offers a 'Hothouse' of ideas and follows the double diamond design process: *define*, *discover*, *develop* and *deliver* (Fig. 9.1) (Hunter 2015). The process is characterised by a number of key elements that arguably set it apart from other problem-solving processes:

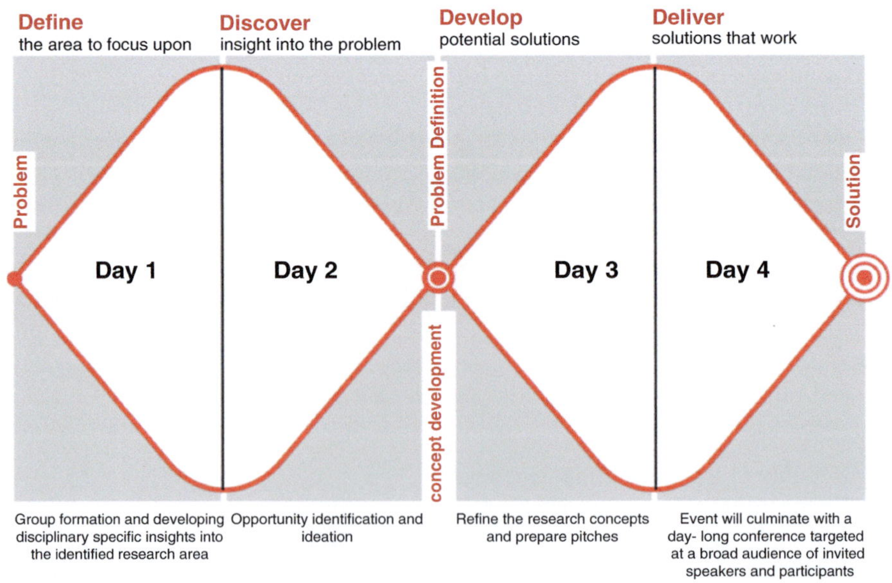

Fig. 9.1 The double diamond of design as applied to HRFL

- The main focus is on developing deep understandings of people, places, situations and contexts. Through understanding the lived experiences of people, we can develop solutions that meet real needs and potentially improve the quality of people's lives.
- This is achieved through embodying ideas through prototyping and then through the testing and iterative development of these ideas. In this way, they lead to more and better ideas.
- The process moves from phases of definition and clarity to ambiguity and openness. Those operating within this process must be open to change, comfortable with uncertainty as the process itself is nonlinear and in continuous flux.
- Reflection, critique and constant questioning to ensure all ideas are robustly tested and refined, ideally leading to the emergence of one or a number of solutions that best address the challenges under exploration.

In building on the previous experiences, the organising group adapted this process to work in the context of research being undertaken by faculty and research staff at Irish higher education institutions and public/private/non-profit organisations. Indeed during the HRFL itself, it was important to stress to the teams that they were not designing a specific intervention but designing the (research) project that would enable the creation of new knowledge in a particular field. The design process underpinned the entire workshop, but there were a series of exercises interspersed throughout the days. These exercises served to open the participants up, allow them to place their individual practice in the context of the team, encourage collaboration and focus on decision making and moving forward. Cards for these exercises were included in each individual participant kit (Fig. 9.2). Having tangible cards served as prompts and guides allowing the teams to customise the process to suit their pace, skills and needs.

A context mapping exercise launched day one's workshops (Visser et al. 2005). Here the individuals mapped their work outlining the *where*, *with whom*, *with what*, *why* and *how* of what their current work entails. They then placed it in the context of their teammates using the cues of *overlap, missing, connected, disconnected, differences, hidden, similarities, compare, contrast, gaps, apparent, new and existing*. Through this the individuals listed the skills and strengths within the team and potential crossover areas and highlighted any research spaces in which they could work.

The second set of cards, contained in the kit, were design tools to guide the teams through the process of the double diamond. The cards were divided into four groups and explained a range of techniques the teams could use to work through the process:

- **Discover:** The process of teams discovering, uncovering and understanding in an effort to identify needs and opportunity spaces. *Empathy Tools, State of the Art, Forecasting, User Research, Storytelling*
- **Ideation:** Ideation is the part of the process that aims to harness the collective creativity of the team to explore, innovate and generate a wide variety of solutions to their opportunity spaces. *Provocation, Brainstorming, 'How might we..?', Backcasting, Speculation*

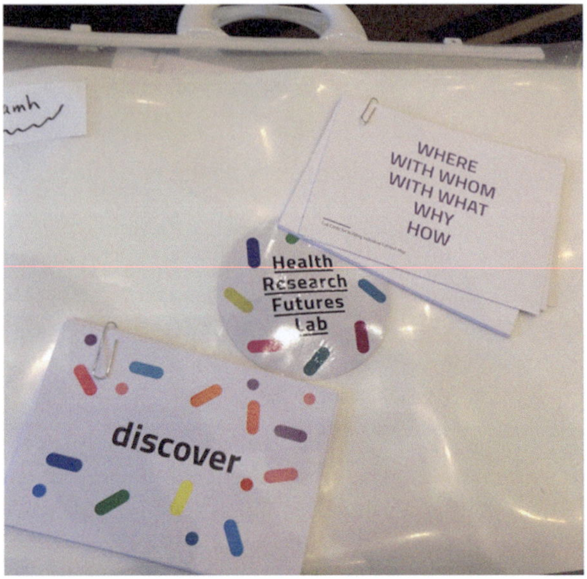

Fig. 9.2 Each participant was given a 'HRFL pack'/kit which included copies of design tools as individual cards.

- **Selection:** Following the collation of all the ideas generated through the ideation process, they must be organised and filtered to allow the most innovative, exciting and appropriate concepts to emerge. *Filtering, Organising Themes, Relevance, Importance, Networks, Scale, Measurable*
- **Prototyping:** Playing out/building/working through an idea is excellent for communicating a concept, garnering feedback and refining features. *Concept Mapping, Storyboards, Scenario Testing, Making*

Participants were asked to share previous experiences of using these techniques or similar ones in order to make the tools more relevant to their practice and to include a variety of techniques/tools from other disciplines.

The importance of guiding the teams through a facilitated process supported by tools cannot be underestimated. The process is often complex and ill-defined, and providing structure, support and anchor points helps the teams to navigate the complexity and make decisions at key stages in this process in order to progress. Below is an overview of the final projects developed by the newly formed interdisciplinary research teams (Table 9.1).

Overview of Interventions and the Design Team's Research Process

Throughout the HRFL, participants were asked to answer a series of qualitative, reflective questions (see examples below). This was an anonymous exercise, where participants were free to be honest and reflective of their experiences. The design

Table 9.1 Overview of the final projects developed by the newly formed interdisciplinary research teams

	Overall area of proposed research	Role of interdisciplinarity	Disciplines represented in the team
Team 1	Antenatal care	Whole person perspective, holistic view of '1000 day' health and well-being for mother and child	Psychology, art therapy, dance, public health
Team 2	Physical activity (PA) for adults living with non-communicable diseases (NCD)	To consider how arts and culture can enhance levels experience of PA through reinterpretation of everyday tasks and environments	Physiotherapy, visual arts, public health
Team 3	Experience of chronic pain	To understand how the individual understands and communicates their experience of chronic pain utilising creative and narrative approaches	Dance, visual art, medical humanities, design, humanitarian action
Team 4	Informal care	To examine informal care identity though discourse analysis	Computer science, psychology, English
Team 5	Diet and lifestyle	To facilitate and examine intergenerational food experiences using creative practice	Visual arts, mental health, management, dance and chemistry

team was mindful not to trigger participants to be evaluative of the HRFL experiences in terms of quality of food or whether or not they 'liked' it or not. Instead we wanted participants to give an honest account of their journey through the HRFL, the ups and downs and key learning moments.

While the HRFL participants were primarily concerned with the development of specific projects within interdisciplinary teams, the HRFL organising group also wished to observe and record the practice of interdisciplinary research, focusing on the lived experience of those involved. As Helen Phelan (2012) writes, 'the emergence of "practice" as a site and source of knowledge is a key development within a range of disciplines including philosophy, sociology, anthropology and cultural theory'. Given the emphasis on relational engagement, negotiation and co-production within interdisciplinary research, an emphasis on 'practice' as a primary field is appropriate. Phelan (2012, pg. 69) continues that practice theory considers subjects as 'social actors produced by contextualized practice and equally as agents in the construction of that practice'. In order to explore the participants' experience of the HRFL as a practice, we designed feedback tools that adapted the strategies and research methodologies used within arts practice research for the purposes of the HRFL. The adaptation of arts practice research methodologies is appropriate in the case of interdisciplinary research, given their emphasis on engaging with creative processes, on capturing data while engaging in practice and on the practitioner as a core source of knowledge (Bell, 2014). These feedback tools therefore facilitated the research objectives of the HRFL organising group, which were to understand the contexts, processes and lived experiences of interdisciplinary research.

The participants were introduced to the idea of this feedback mechanism at the beginning of the first day and advised that they were under no obligation to engage with this, and nonengagement would not have an impact on their overall participation in the HRFL process. Following this, consent forms were circulated. All participants agreed to be part of this research process. Each participant was provided with a feedback pack, which contained the following reflection sheets (akin to a qualitative questionnaire), and reflective journals (with room for open writing) for each of the 4 days of the HRFL.

The reflection sheets consisted of a series of questions, designed to prompt reflections on process and interaction, rather than on the projects underway. The participants were advised to take some time to fill in the questions during the day or during the afternoon and evening breaks. All feedback forms were anonymized, but the sheets in each pack were identified by a number, allowing us to collate the responses of individual participants at the end of the process. The questions focused on the individual experience, on individual contribution to group dynamics and on personal perspectives on taking part. For example:

- *Describe any working patterns that were new or challenging to you today.*
- *Did you feel comfortable taking part in these? If so, why? If not, describe what made you feel uncomfortable.*
- *Do you feel confident that your group would be able to progress a research project proposal in the future? If so, why? If not, describe why you do not feel confident about this.*

We gave participants frequent reminders throughout the HRFL to ensure that they did engage with this as they worked, rather than filling them in at the end. This we felt would negate the opportunity to observe 'real-time' practice and changes in attitudes towards interdisciplinary work processes. Given the short time frame of the HRFL and the desire of the participants to maintain focus on their project development, there were some tensions surrounding the completion of the reflection sheets, but most participants engaged with it in a very robust way, and the resulting data is very rich. The reflective journal entries, which were to be completed in the period between the two HRFL sessions, were less successful, with minimal take-up from participants. Similarly, there was minimal engagement with the final feedback task, which asked participants to submit an audio recording of a 10-min interview between themselves and a colleague from their own disciplinary background about their experience of the HRFL. This tool was designed to gain insight into the way in which an interdisciplinary experience was narrated back within a specific disciplinary environment. For example: *Do you think that you are more or less likely to develop interdisciplinary research approaches within your future work?*

The HRFL group's experience of conducting research into practice within the context of interdisciplinary research reflects the need to gather reflective feedback during the work process, rather than asking individual participants to engage with this beyond the HRFL itself. The intensely focused environment of the HRFL allowed the group to gather very rich data from participants, but this focus was not maintained beyond the structured environment of the HRFL itself. In the following section, we draw on this data to give an account of researchers experience of the HRFL and the nature and role of experiential exercises specifically.

Use and Role of Experiential Exercises

During the course of the HRFL, two experiential workshops were introduced. The main aim of these was to enable trust and the formation of community (Moore, 2003). They focused on the body and connecting with others through the body. The exercises were simple, with the purposes of lowering 'fences' and 'levelling out' the group and overcoming possible stereotypes held with regard to people being 'artists', a 'doctor', 'PhD researcher' and so forth. In order to create an environment where people would see and listen to each other equitably, the exercises were devised to bring the attention to the sensorial body and to transform concepts into embodied facts. For a couple of hours, the practice was not entered through knowledge carried by language skills and the individual background of each participant was not foregrounded. As a result people loosened up their boundaries and experienced the exercise, if one can say, 'at the same level'.

The diversity of games and exercises went from individual to group. The first workshop brought the individual to the centre of the exploration. The exercises emphasised aspects known to channel contact with individual perceptions, such as the breathing cycle. When the individual brings the attention to a micro-space such as breathing, the mind is put at ease, thus restoring a sense of integrity and self. The participant is invited to focus on the present time, avoiding the stress of the next stage and of previous experiences.

Listening and observing are some of the key elements in interdisciplinary work or any other collaborative process. Listening to our own breathing is an effective way to connect to something real and vital. This simple exercise becomes a tool that the participant can use at any time to change focus to the self and allow to engage the tasks of connecting with others with less stress.

> Stopping to centre oneself (in terms of relaxation, breathing, body awareness) is welcome, as part of a day of thinking. Participant D

The participants also engaged in exercises/games that were purposely devised to confront them with feelings such as fear, trust and instability. The following exercises guided the participants to explore the sense of shared responsibility and teamwork. The physical and concrete perception of internal/external limits and borders helped the participant to better understand his/her own restrictions, to (in most cases) softly overcome them.

> The sessions with partners on trust, relaxation etc. opened my mind to the need of build [sic] more mindfulness techniques into my work. Some were difficult because we were working with strangers but this was overcome very quickly. Participant O

Some of the proposed games required working with someone else, thus creating a spontaneous relationship. The exercises were repeated a number of times and lasted long enough to allow, at different stages, the confrontation with feelings such as fear and mistrust. The individual recognises a learning pattern, which is internalised, enabling the participant to relax and ease into it. The physical experience teaches the body, the body eases the mind, which in turn understands that there is nothing to fear.

> [the exercises] …were invaluable in dispelling tension and in promoting collaborative spirit. The discussion in my group became easier after the exercise. Participant E

The participants laughed together, breathed together and moved together. Such actions brought a sense of community in which the different people, status and backgrounds started to all work towards a shared purpose. The workshops intended to reveal that the embodied encounter could prove beneficial to narrow the gap between the different universes and create a particular state of mind that facilitates interdisciplinary interaction. The success of that small experiment may lie in the increased acquaintance amongst participants – but also in the inhibition of behaviours and reactions, which, while they may define the form and function of a field, may render its interaction with other fields difficult. The interruption of an empirical body practise into interdisciplinary interaction of academic nature suggests the necessity to allow initiating the encounter at a level that is neither rational nor conceptual, hereby levelling the systems of values and acknowledgement at play within each field.

Role of Research Support Staff

A further dimension of the HRFL that is worth considering is the role given over to research support staff. As part of the design of the HRFL, we believed that disciplinary-specific research support staff should be involved, in two key ways: firstly, to observe the unfolding design of new research projects and, secondly, to offer specific advice to the newly formed teams in clinics which were a focus in particular of day 3 of the workshop. Therefore, over the first 2 days, research officers were invited to spend time in observation mode. Research officers in attendance included a national contact point for Horizon 2020, University level AHSS research support as well as school-level support staff. The reaction from these personnel to the experience came as a surprise. While for the organising group, it seemed self-evident that research support staff should be on hand to guide the research teams to possible sources of funding, for the staff themselves, this proved to be a unique opportunity to see research in development stage. What was more usual, they suggested, was to be brought into advice on a project very close to a funding round deadline. Throughout the HRFL they were able to offer insight from their experiences of reading final research proposals, get a richer understanding of the iterations involved in new research project development and in effect steer the direction of the research in light of funding opportunities available. During the clinic research on day 3, the research support staff had a formal opportunity to offer advice on shaping the research proposal to be legible by possible funders. These clinics also included the participation of the local arts officer to provide suggestions with regard to arts-specific funding opportunities. There were also two further clinics with regard to pitching techniques and another one relating to interdisciplinary work.

Responding to Emerging Issues in Interdisciplinary Research

In this section we reflect on the lessons and insights for interdisciplinary research emerging from the HRFL. Due to the nature of the lab and the overall design approach, one barrier to interdisciplinary research became particularly observable, that is, the differing values that researchers/practitioners bring to their work and their understandings of the legitimacy of their knowledge base in relation to their new colleagues. Initial reflection on the questionnaire data submitted throughout the research process indicated some frustration within research groups based on differing methodologies and approaches. This was also evident in the design team's observation of the research groups towards the end of the second day. In order to address this potential barrier or block its development, we devised a short workshop based on the idea of value in/of research.

The design team's observation was that the frustrations arose due to a clash in value systems around the validation of research. Validation systems and cultures of validation differ widely across different disciplines, and a failure to address this could lead to the implementation or domination of one method over others (i.e. 'hard' research data or experiment results vs. practice-based research or critical discourse analysis). The short workshop involved researchers working in pairs and discussing a series of questions about what they value in their own research process, where they learned to value it in that way, and what potential blind spots might exist within that system. This reflective approach was loosely based on the d.school design thinking model (in terms of creating user empathy) and also asked researchers to explore the fact that their systems of value were learned, rather than innate, and to consider the blind spots within them themselves rather than having them pointed out in a potentially confrontational situation, which would lead to a defensive rather than reflective approach.

The following are some quotes from participants regarding their overall experience of the HRFL. These are from unsolicited emails sent following the workshop:

- *I am now ready to email you and express my appreciation and gratitude for your encouragement, light heartedness and expertise over the Research forum experience. Now that I'm on the other side with a nice gap to regulate myself again, I feel very appreciative that I had the opportunity to meet so many researchers and people who embodied deep curiosities and passions for a wide field of interests and disciplines. I did find the 2nd two days a bit of [sic] whirlwind and perhaps was not fully prepared in myself for the intensity and heat in striving towards the end goal of the presentation and proposal. But I did really appreciate the fact that I didn't give up and stuck in there excavating knowledge and ideas and working together. I have to say that I felt very supported at all times by your holding of the group and what I saw as a gentle yet firm skill in navigating the unknown.* HRFL participant. Dance Professional

- *I just wanted to send a heartfelt thank you to you all for the care and expertise you brought to the Health Future Research* HRFL. *It was incredibly challenging at times and I even gave Annemarie a post-it with the sentence "To never see anyone on my team again!" in response to the question" what would we like to see your team do next with the project" (I am NOT melodramatic). However having spoken to my team and sticking with it I feel I learnt in four days what would probably take me a few years to learn otherwise. The underlying feeling was that we had a cushion of support, time and space to really expand beyond our comfort zones of research and inhabit new ways of creating.* HRFL participant, academic/performing artist
- What will you 'take away from the workshop': "*I will take a developed understanding of what the "arts and culture" is, plus how it can contribute to solving problems in the medical field, from talking to others not in my group. I also learned of other aspects beyond arts/culture, e.g. humanities, language.*" HRFL participant, Public Health Academic
- *Working process was refreshing in some ways, in that sharp deadlines brought focus – not without its challenges. The design tools offered through the course gave a step-by-step guide that I would like to use again.* HRFL (anonymous)

The feedback obtained reflected positively but also honestly on the difficulty in putting together not only different disciplines but different personalities. In the following we offer an overview of more general lessons to be learned from attempting to bring together new research teams across disciplinary divides.

It is important to develop ways to:

- Overcome language as a barrier and recognise that a shared language system is an outcome of interdisciplinary working and not an input.
- Respond fluidly to the changing needs of the teams. Rigid processes do not suit interdisciplinary teams as they form. Process needs to be responsive, iterative and evolving.
- Tools and supports are useful to help overcome some of the challenges and help move the team through their process. We have summarised/listed some of the tools that can be included here.
- Time frames are important – teams feel motivated as they move towards deadlines, helps make decisions and reach consensus.

The design team would also like to offer the following insights that underpinned the design of the HRFL and are key to enabling interdisciplinary research and especially to bring together researchers and practitioners from STEM and AHSS.

For Early Career Researchers

- Benefits of facilitating (enabling and equipping) interdisciplinary working and not just workshops that talk about the topic.
- Giving these researchers opportunities to establishing networks. Identify their areas of focus.
- Lower confidence – there are greater opportunities for mentorship.

For Senior Researchers

- Offering SRs opportunities to share their experience but in a co-working environment, i.e. not just to talk about their prior experiences but integrate these into new mixed interdisciplinary teams.
- Offer SRs opportunities to broaden their research area and explore new methods and approaches.
- Building an environment where SRs and ECRs can develop respect for the approaches of other disciplines, the languages used and the contributions each discipline can bring.

Supervisors/Mentors

- Explore diversity of methodologies and approaches offered by the different disciplines.

Recognise the struggles that will exist and develop as teams form. Supervisors need to be equipped with the capacity to deal with and facilitate the overcoming of these challenges.

Funding Agencies

- Involve disciplines that might not be traditionally involved in research (e.g. visual and performing arts).
- Expand and broaden their remit to allow new exciting areas of research to emerge.

Institutions

Explore potential for innovative and novel research areas and expertise to emerge. Meet the requirements of funding agencies such as H2020 for interdisciplinary research. A key recommendation from this workshop is for institutions and research funders to give attention to interdisciplinary *codesign* of research projects.

Final Thoughts

Central to the design and implementation of the HRFL was the notion that a key aspect for researchers is the importance to literally *practice interdisciplinary research*. This is based on the premise that interdisciplinary research is not self-evident and is quite challenging. The challenges can be profound in terms of the values researchers hold as well as their ontological and epistemological commitments. But these can also be more mundane and involve different work practices and routines making the practice of bringing together research groups from across disciplines logistically difficult. It is Important, of course, to acknowledging *differences* between disciplines, but *not to overstate* these. Allowing teams to spend time with one another can open up understandings of what they have in common. We need to encourage mix method approaches and be more aware of single discipline dominance in work packages.

In the HRFL we worked hard to integrate *research officers* throughout the process. This was important for them and for the researchers. For them, it gave them an insight into how a research project develops and to witness the challenges of interdisciplinary research team formation. For the researchers, hearing about the research funding process throughout the time when they developed their research kept them on track. Finally we would say that *showcasing* interdisciplinary best practice across different research themes is important, in order to present different experiences and examples to Irish and indeed worldwide researchers.

References

Bell J (2014) Doing your research project: a guide for first-time researchers. McGraw-Hill Education, Maidenhead

Hunter M (2015) The design process: what is the double diamond. Design Council, London

Lyall C, Bruce A, Tait J, Meagher L (2015) Interdisciplinary research journeys: practical strategies for capturing creativity. Bloomsbury Publishing, London

Moore S (2003) Writers' retreats for academics: exploring and increasing the motivation to write. J Furth High Educ 27(3):333–342

O'Malley L, Ryan A (2006) Pedagogy and relationship marketing: opportunities for frame restructuring using African drumming. J Mark Manag 22(1–2):195–214

Phelan H (2012) IMBAS: performing a philosophy of music education. In: Bowman W, Fregga A (eds) Oxford handbook of philosophy in music education. Oxford University Press, Oxford

Visser FS, Stappers PJ, Van der Lugt R, Sanders EBN (2005) Contextmapping: experiences from practice. CoDesign 1(2):119–149

Chapter 10
Internationalisation in Higher Education as a Catalyst to STEAM

Armida de la Garza

Abstract In as much as academic disciplines can be regarded as cultures in their own right, the internationalisation and the interdisciplinarity agendas largely overlap: both seek to introduce students to knowledge of other cultures and intercultural competence; and both seek to instil global perspectives in students, so they can acknowledge our common human destiny. As Internationalisation efforts in Higher Education are usually led by the institutions' International Offices in partnership with the academic units at various levels, they are particularly well placed to promote interdisciplinary collaboration. This chapter discusses ways in which Higher Education institutions can take advantage of these synergies to cultivate and nurture STEAM through internationalisation.

Keywords Internationalisation in HE · Internationalisation at home · Internationalisation of the curriculum · Interdisciplinarity

Introduction

Ours are times of paradox and change. We have reached a stage in which science and technology have on the one hand enabled us to live longer, provided us with instant global communication and revolutionised production to include robotics and automation—and with this, the possibility of a radically different future. But on the other hand, the rapid increase in population and the same positive developments just outlined mean this is the first human generation whose actions will flood low-lying islands, whose rate of resource consumption is well above two and a half times the production capacity of the planet and whose supply of drinkable water and clean air to breath are not guaranteed (Brown et al. 2010, 3–5). Issues such as these and climate change, global health, urban violence or coping with biodiversity loss are all examples of 'wicked problems', that is, problems for which there can be no final solution since they are part of the society that generates them, and any changes to

A. de la Garza (✉)
College of Arts, Celtic Studies and Social Sciences, University College Cork, Cork, Ireland
e-mail: adelagarza@ucc.ie

© Springer International Publishing AG, part of Springer Nature 2019
A. de la Garza, C. Travis (eds.), *The STEAM Revolution*,
https://doi.org/10.1007/978-3-319-89818-6_10

the situation introduce further issues. Our incapacity to address wicked problems has been traced to the compartmentalization of scientific and professional knowledge, to the sector-based division of responsibility in contemporary society and to the increasingly diverse nature of the social contexts in which people now live (Lawrence 2010, 16). Transdisciplinary research[1] and practices offer an avenue for the STEM disciplines, the arts, humanities and social sciences (STEAM) to overcome these obstacles and tackle these truly vital issues. It also introduces a model of accountability to society and promotes innovation as previously separate fields are brought in contact with one another. Transdisciplinary models of knowledge production are a necessary response to demands that academic life becomes more integrated with society and the economy. At any rate, it is estimated that each day we generate the knowledge contained in all the world's libraries, so transdisciplinarity may well be the only sustainable way forward as regards the production and application of knowledge: 'the world is more ecological than we had thought [...] rather than it being possible to study phenomena in isolation, everything is implicated with everything else [...] life is lived in media res' (Frodeman et al. 2012, xxx–xxxiv).

Importantly, all the wicked problems mentioned above are global in nature and require a global citizenship mind-set to be addressed. But culturally, the world has become increasingly more divided as it has become more connected, and there is an urgent need to build bridges of tolerance and respect through education. International interaction and collaboration through education can foster cross-cultural insight and exchange that is enriching and enabling 'for individuals, communities, nations and the world' (Leask 2015, 17) and potentially as useful to pool towards the resolution of our most pressing issues as the disciplinary knowledge. Universities have attempted to face this challenge with internationalisation of the curriculum (IoC) and internationalisation at home (IaH) initiatives. Briefly, IoC entails the incorporation of 'an intercultural dimension into the content of the curriculum as well as the teaching and learning processes and support services of a programme of study' (Leask 2015), while IaH, a subset, is defined as 'any international activity with the exception of outbound student and staff mobility' (Green and Whitsed 2015, 7).

Thus transdisciplinarity, especially in its STEAM variety, seeks to respond to our common human challenges through the integration of science, arts and humanities. And internationalisation efforts seek to provide a fuller, more rounded education to individuals by integrating knowledge produced in different cultures. Both approaches share the need to transcend cultural boundaries, the one from disciplines, the other from national cultures and both from the institutional contexts in

[1]Although it is common to understand multidisciplinary research as that which draws from various disciplines, each bringing a contribution towards the resolution of a problem, interdisciplinary research as that in which synthesis occurs and transdisciplinary research as that which frames the question 'as part of a total system without any firm boundaries between the disciplines' (Barry and Born 2013, p9), here I take the term to mean the collective understanding of an issue created 'by including the personal, the local and the strategic, as well as specialised contributions to knowledge' (Brown et al. 2010 p. 4), which goes beyond academic knowledge.

which higher education is embedded. Importantly, both ascribe a central role to imagination, normally associated primarily with the arts. Transdisciplinary perspectives recognise that 'leaps of creativity in the formalised science are exemplified in the classification of life forms by Linnaeus, which differed significantly from those of Aristotle before him and Darwin after him [...] For each inquiry across the ages, a flight of the imagination led to fresh scientific concepts and images which changed the interpretation of reality' (Brown et al. 2010, 9), whereas the key difference between the internationalisation of the curriculum process and commonly used curriculum review cycles is 'stage two, the imagine stage. It is essential and integral. It stimulates creative uncertainty through challenging the traditional and taken for granted [...] and inviting broadening and deepening of engagement with difference' (Leask 2015, 42). And notably, both entail considerable personal, even transformational, change. STEAM values the capacity to be flexible and adapt in teaching and research: 'under the shadow of the mindless repetition of old lectures and the artificial extension of exhausted research programmes' the ability to undertake transdisciplinary research 'is seen as a mark of flexibility and adaptiveness, which are highly valued in today's knowledge economy' (Graff 2015, 2). From the perspective of graduate employability, it has even been argued that the 'skills' framework of reference that has been so central in the literature be modified for a focus, after Pierre Bourdieu, on capitals: human capital (the hard skills and technical knowledge), social capital (that which creates a bridge between graduates' educational experience and helps them broker their access to job openings), identity capital (when graduates harness their sense of personal identity around targeted employment, including channelling existing lifestyle and extracurricular activities towards that goal), and psychological capital, i.e. levels of resilience and adaptability (Tomlinson 2017, 18), very highly valued and promoted by holistic approaches such as STEAM. And IoC theories speak of the internationalisation of the academic self. Understanding one's own linguistic, sociocultural, political, ethical and educational constructs, values and beliefs, as well as the process whereby they were formed though enculturation 'has continuous relevance in the ongoing project of intercultural teaching and learning across curriculum that aspires to "internationalisation"' (Green and Whitsed 2015, 10). As put by Green and Whitsed, if the ideal graduate is interculturally sensitive and competent, a socially responsible and globally aware citizen, then the ideal lecturer in the present context is one that broadens curricula and incorporates pedagogical approaches (p. 10).

However, despite all this fundamental common ground between STEAM approaches and IoC and their radical ambitions, so far IoC has been pursued exclusively around disciplines. This may be because like the national cultures that IoC seeks to draw from, the disciplines have been perceived as different cultures, at best as 'separate communities of practice with their own organisations, power hierarchies, questions to answer and [sometimes heavily policed] entry boundaries' (Brown and Harris 2014, 115), if not as 'artificial holding patterns of enquiry' with metaphysical significance that cannot be overestimated, according to Steve Fuller (quoted in Graff 2). Still, IoC takes it as given that:

> Disciplinarity exerts enormous power and influence over the organization and production of knowledge and discipline groups are global communities […] Discipline communities transcend national boundaries. They provide an organizational focus for universities and the curriculum across the world. They are at the heart of the process of IoC. (Leask 2013, 99)

In the paragraphs that follow, I propose an alternative view of IoC, as an ideal enterprise to pursue STEAM agendas, for as I have shown the two in fact overlap to a large extent. Inasmuch as IoC seeks to develop students' international and intercultural perspectives as global professionals and citizens, it calls for engagement with the arts, humanities, social sciences and sustainability initiatives across programmes. Moreover, even the 'hardest'[2] and purest natural sciences are taught, and teaching is always a socially constructed activity (Carroll 2015, 104), thus any IoC in science must by necessity engage education and other humanities. Where relevant I illustrate with examples from the experience with IoC and STEAM at University College Cork, Ireland.

Internationalising the Curriculum for STEAM

Essentially, internationalising the curriculum is a form of critical participatory action research in which teams of academics responsible for the curriculum at the programme—or occasionally course—level actively enquire into their own teaching practice and their students' learning process to inform their understanding and make improvement in order to achieve the following learning outcomes:

1. Global perspectives. IoC demands knowledge of other countries and cultures and competence in other languages. One way to promote STEAM would be to include computer programming languages in the options available to all students (Denicolo 2013, 53).
2. Intercultural competence. Sensitivity to the perspectives of others and a willingness to put oneself in their shoes. An understanding of the nature of racism. This intercultural competence can also be promoted by requiring students to frame a problem from an arts and humanities or scientific perspective accordingly.
3. Responsible global citizenship. Understanding the necessity to engage with sustainability, equity and social justice. This learning outcome is a shared goal of STEAM.

IoC considers the curriculum both as formal, comprising the syllabus as well as the organised experiences that are part of a student's programme of study, and informal, consisting of the support services and additional activities available on campus. In addition, the 'hidden curriculum' is taken to mean the unintended and

[2] A much utilised taxonomy of knowledge that classifies disciplines as hard science (mathematics, physics), hard applied (medicine, civil engineering, pharmacy etc.), soft applied (psychology, law, business and economics) and soft (art, design, history, media studies) has found the last three far more open to IoC.

implicit messages of whose knowledge is valued and indeed what counts as knowledge, since every selection implies an omission—for instance, although the IoC perspective considers indigenous knowledge, in contexts in which science is taught, art can often be perceived as knowledge that is less valued. Together they make up the total student experience, and rather than being separate and discrete, the three overlap to some extent. At all these levels, and in particular in the area where they overlap, opportunities are provided for interventions that seek to attain the above learning outcomes.

As mentioned above, the IoC framework puts the disciplines at heart since they define the scope of the curriculum. They are constrained by institutional, local, national and global contexts and must engage dominant and emergent paradigms. Requirements of professionalism and practice, assessment and the need for systemic development also contextualise and constrain the curriculum (Leask 2015, 27). But to internationalise the curriculum for STEAM, recent changes in the role of the disciplines in universities must be taken into consideration. Paul Trowler contends disciplines are undergoing significant transformations due to the impact of global and domestic market forces, casualization of the academic workforce, the amalgamation of departments into single units or their closure (Trowler 2012). Wendy Green and Craig Whitsed also note the growth of interdisciplinary 'domain based' studies, such as 'women's studies' or 'environmental studies', resulting in disciplines being reconfigured as spaces of polyvocality where multiple, conflicting narratives co-exist (2015 280). Further, they offer a useful metaphor to conceptualise this change in which the former regime of production of academic knowledge in universities could be described as analogous to a chess game: a closed space of territorialisation with each piece, the disciplines, coded with a pre-determined hierarchy and organised according to its function. The present regime is by contrast similar to a Go! game in which pellets are situationally defined, movement is relatively free and pieces operate in open space where power is fluid. As the game is being played, the identity of any given disc changes in relation to other discs, so they are always in a process of becoming (p. 281–2). Indeed, their research entitled 'Critical Perspectives on Internationalising the Curriculum in the Disciplines' actually reports numerous similarities in the challenges faced by a variety of disciplines, while also intradisciplinary differences. They conclude that IoC 'necessitates the development of critical interdisciplinary spaces which foster the exchange of innovative ideas' (p. 280). By the same token, Harvey Graff notes that we easily assume differences between disciplines and interdisciplines rather than relationships and connections and that the focus tends to be on the addition of disciplines rather than their interactions (Graff 2015, 6). Even Betty Leask, who at one point characterises the disciplines as 'the life-blood of higher education, providing both an organisational focus for the university and the curriculum and a social framework' (p. 28), later admits that 'it can be an advantage to encourage interdisciplinary conversations and debates' while internationalising the curriculum as 'this can be an effective way to stop the censorship that is often practiced by discipline communities on their colleagues' (p. 110). In sum, from epistemological reasons on the nature of knowledge in the present context to the practical business of discussing how to internationalise the curriculum in a given programme, a transdisciplinary approach that puts STEAM at its heart is better suited and can be more effective to pursue IoC.

IoC should be a volunteered process, undertaken with an open mind and in diverse groups. The planning team thus bring their disciplinary strengths, past experiences, cultural backgrounds and skills, all of which are also useful to consider, including a transdisciplinary outlook. The process starts by identifying and seeking to maximise programme-level opportunities and benefits. As all sound policy design, it should look forward and prioritise values and outcomes, aiming for transformation: of the curriculum, concepts, students and ultimately university communities (Carroll 2015, 105). It is iterative, linking together decisions about design, teaching and assessment and making connections between them explicit. It consists of five steps, namely, review and reflection, imagining, revision and planning, action and evaluation. The questionnaire on internationalisation of the curriculum or QIC is an often-employed tool, which stimulates reflection and guides the discussion. Below I list the main steps drawing from Leask (2015, 41–50) adding the way in which the same step can be used to embed STEAM in the curriculum at the same time (Table 10.1).

The STEAM-IoC Synergies

Synergies between the two agendas are evident at various levels. First of all, working from the learning outcomes of an IoC, namely, global perspectives: IoC demands competence in other languages. Disciplines have often been regarded as entailing specific 'languages', with attempts at cross-disciplinary collaboration often thwarted by colleagues 'not speaking the same language'. Indeed, the whole inter-/transdisciplinary enterprise is sometimes conceptualised as a form of translation from one discipline into another. Having defined translation[3] as 'an act of invention that works by combining different elements into a congruent whole', Michel Serres has theorised that art can often be regarded a translation of science and vice versa, if translation is regarded a process of communication that entails making connections and forging spaces between different domains. To Serres, these passages 'have the power to distort and transform' (Guldin 2016, 111). He further defined science as the sum of all messages optimally invariant with regard to any translation strategy, and persuasively argued Leibnitz has been translated into mathematics and Blaise Pascal into the paintings of Georges de la Tour and that William Turner translated thermodynamics physics into his painting *The*

[3] Indeed, in science, metaphors, whether exegetical/pedagogical or theory-constitutive, are essential. They generate insight and help in perceiving connections 'that once perceived are truly present. They enable us to see aspects of reality that the metaphor helps to constitute. Metaphors show that conceptual boundaries are elastic and permeable and can be stretched and altered' (Guldin 2016 13–14), demonstrating the value of linguistic approaches to students of science.

Table 10.1 The IoC process (Leask 2015) revised for STEAM

Step	For IoC	For STEAM
Review and reflect	Establishing a rationale: What international/intercultural knowledge, skills and attitudes will students need as graduates of the programme?	Establishing a rationale: What interdisciplinary knowledge, skills and attitudes will students need as graduates?
	Review content, teaching and learning arrangements and assessment, identifying opportunities to develop the skills/attitudes or introduced the knowledge deemed as essential	
	Reviewing student feedback for strengths and weaknesses in relation to internationalisation	Reviewing student feedback for strengths and weaknesses in relation to inter/transdisciplinarity
	Comparing and contrasting feedback from international students	Comparing and contrasting feedback from students from other programmes in different colleges
	Reviewing feedback of other stakeholders, such as industry partners and professional associations	
	Reviewing institutional goals in relation to internationalisation	Reviewing institutional goals in relation to inter/transdisciplinarity
	Reflecting on achievement and identifying opportunities for improvement	
Imagine	Discussing the cultural foundations of dominant paradigms in the disciplines and their relation to the curriculum	
	Identifying emergent paradigms and considering the possibilities they offer	
	Imagine the world of the future, including what will be needed to work effectively and ethically then	
	Imagine different ways of doing things	
	Brainstorm a range of possibilities to deepen and extend internationalisation and inter/transdisciplinary approaches in the curriculum	
Revise and plan	Establish programme-specific goals for IoC	Establish STEAM goals to be included
	Detail intended learning outcomes and map the development and assessment in the programme	
	Identify blockers and enablers, experts, champions and resources	
	Set priorities and discuss how the changes will be evaluated	
Act	Implement new teaching arrangements and if necessary support services. Include workshops and, as required, new assessment, new units, courses, electives or rubrics	
	Collect evidence required for the evaluation of changes	
Evaluate	Analyse evidence, reflect on impact, consider gaps, summarise achievements	

Fighting Temeraire (1838). For Serres, not only is the painting depicting the homonymous old sailing ship being towed by a combustion engine tugboat towards the place where it will be broken up for scrap a scale model of what is happening in society, with the engine replacing the sail, but also 'the canvas spit flames like steamboats. It stages Carnot's reflections on heat and temperature and their relation to energy and work [...] Turner anticipates the future theoretical developments of physics without having any direct knowledge of Carnot's thermodynamic circle' (quoted in Guldin p. 113).

In the same vein albeit from the perspective of physics, Michael Leyton has argued that inasmuch as shape is the result of force applied to physical objects, it can be regarded a means whereby the transformations that led to a given state of being are stored, or in his words, 'shape is equivalent to memory storage' (2006, 1). Thus the shape of a bay for instance results from an inflow of water at the top part of a coastline that starts to dip inwards until a resistance against the water flowing is met. In this sense, the shape of the bay stores the memory of the process whereby it was formed. Building on this insight, Leyton has developed a theory on the appreciation of painting based mainly on mathematical and physical criteria—i.e. on mathematical language—paying attention to shape, tension and balance, where 'geometry becomes equivalent to aesthetics' (p. 1).

From these perspectives, introducing students to other STEAM disciplines enables them to learn another language, and the works of artists and scientist teams such as that of Anna Dumitriu and her physician collaborators, featured here in **Chapter 7** as an example of art that is created by the means of bacteria, or of engineers and artists who design robots for the aesthetic beauty of their movements (Herath et al. 2016) are instances of languages in translation. The benefits of fluency in the 'languages' of both drawing and science can be expressed thus:

> The act of drawing is an act of recording. Science requires the recording of data to seek insights and patterns […] Long before the technologies of printing, photography, and digital imaging, drawing was the only way to create a representation of features, construction, orientation, or pattern [...] The alphabets and numerals that were developed later are drawn symbols for already standardized language. In mathematics, the symbols combine as visual shortcuts of communication to describe relationships and patterns. Drawing has been essential to our intellectual development. Where would science be without the drawings of Copernicus, Da Vinci, Audubon, Darwin, Bohr, Watson, Crick and Franklin, in depicting models, processes, and possibilities? (Katz 2016)

Equally, all arts and humanities programmes should include in their reading lists articles from scientific journals and vice versa, to ensure that, as with internationalisation, 'other viewpoints are included and awarded due consideration'. The publication exchange project started at University College Cork in 2017 actively promotes the work of staff members among the university's international partners in close collaboration with Cork Open Research Archive, CORA, the library's institutional repository, which serves as platform for their dissemination and archiving. Funded by the Irish Research Council, the project also receives publication contributions from the partner universities' staff that are interested in working through open access. A digital artist was enlisted to visualise the contents of the works exchanged as well as their spatial dissemination patterns as they are browsed or downloaded abroad, and some abstracts were translated into the participating languages (English, Spanish, Chinese and Portuguese). In this way, not only have the institution's international partnerships been strengthened through this academic collaboration, and its institutional repository has been enhanced, but the project has set the stage for STEAM exchanges when it is expanded beyond its initial scope in the arts, humanities and social sciences.

The second IoC learning outcome, namely, intercultural competence, including sensitivity to the perspectives of others and a willingness to put oneself in their shoes, is currently focused on understanding the nature of racism. This sensitivity to the perspectives of others can be broadened by including gender and understanding of the nature of sexism that is at work in science and the arts as academic fields. Combining and juxtaposing the very different fields of electronics and sewing/embroidering, e-textiles bring together activities traditionally regarded as feminine and masculine and also hand and mind, formal and informal education, visible and invisible technology, physical and digital worlds, low and high tech and the broader meanings of ‘hard’ and ‘soft’ (Buechley et al. 2013). The embroidery has a crucial function conveying electricity, in such a way that ornamental design and functionality are not antagonists. They allow participants to express themselves creatively through the use of technology while developing important new skills, ideas and social connections. Providing digital textile courses and workshops promotes STEAM collaboration at its best while also increasing the number of female students in fields of engineering and electronics. From an IoC perspective, this would provide opportunities for the comparison of male/female roles in different cultures.

Moreover, the third learning outcome, i.e. responsible global citizenship, including understanding the necessity to engage with sustainability, can be addressed through university-wide modules such as the one implemented at University College Cork in 2015–2016, which stands out as a model of good practice. Building on its green campus strategy, the module drew from the expertise of all colleges (Medicine and Health; Business and Law; Arts, Celtic Studies and Social Sciences; and Science, Engineering and Food Science) with volunteer lecturers from each teaching a highly participatory session each week to provide a truly relevant interdisciplinary—from the participants’ perspective—learning experience, for all undergraduate and postgraduate students and staff, as this module was indeed also open to academic and nonacademic staff and eventually to the entire Cork community, including activists, employers and various stakeholders. In fact, following the success of this initiative, a second university-wide module on internationalisation is currently in preparation. In general, university-wide modules address the important issue of institutional blockers to STEAM.

In addition to working through the learning outcomes, a further instance of synergy between STEAM and internationalisation is the IoC process itself. The second step, ‘reviewing and reflecting’ as detailed above, requires identifying opportunities in both the formal and the informal curriculum to include internationalisation actions. Common ways to do this are inviting guest speakers and designing reading lists to ensure international viewpoints are represented and given due consideration: guest lectures by speakers from local cultural groups or international companies, international partner universities or, increasingly, digital learning and online collaboration and comparative international literature are among the instruments to internationalise teaching and learning (Beelen and Jones 2015, 64). These guest lectures or even fellowships can also be recruited for the cultivation of STEAM. For example, the UCC Fulbright Fellowship usually ascribed to the College of Arts,

Celtic Studies and Social Sciences is this time shared with the College of Medicine and Health for a range of activities across the formal and informal curriculum that will develop global medicine, and will be extended to include collaboration with the Environmental Research Institute. In this way, an international fellowship designed to promote scholarly exchange between Ireland and the USA can at the same time promote STEAM.

Internationalisation at Home (IaH)

A subset from IoC, internationalisation at home (IaH) is concerned with extending the benefits of internationalisation to students and staff that do not avail of study abroad or staff exchange programmes. One of the key contributions of IaH lies in framing a context for the development of employability skills (Beelen and Jones 2015, 68). Many studies have shown that international experiences are instrumental in developing the kind of transferable skills that employers value. Certainly, innovation, which employers highly value, is based in creativity, and this in turn is based in exposure to new people and new ideas, 'particularly through transdisciplinary social input' (Bridgstock 2017, 345). Exposure to new people and new ideas can be achieved by actively mobilising for this purpose an institution's network of international students and staff, who are bound to bring with them knowledge of different educational systems and possibly of different paradigms of research and teaching (Altbach and Yudkevich 2017, 2). An employability and transferable skills training programme across disciplines can be matched to these efforts. Such a programme would focus on bringing skills traditionally associated with the arts and humanities—such as aesthetic appreciation, critical thinking or communication skills—to students of technology and science, while also bringing skills traditionally associated with science and technology—such as planning and problem solving, numeracy and the use of information technology—to students of arts and humanities, actively taking advantage of the innovative perspectives that international staff and students bring. Staff mobility is only effective when it is part of a deliberate process of staff development, as noted by Brewer and Leask (2012, p. 251). Likewise, the local community can become the focus of learning opportunities with intercultural and/or international dimensions and expose students to STEAM frameworks.

Conclusions

At one extreme disciplines have been lauded as above 'the life and blood of higher education', while on the other they have been described as an abdication: 'by focusing on standards of excellence internal to a discipline academics have been able to avoid larger responsibilities of how knowledge contributes to the creation of a good and just society' (Frodeman et al. 2012, xxxii–xxxiii). Whichever view one subscribes, there is no doubt that disciplines were central to academic life of the twentieth century. But technology has changed society in a fundamental way. Fast data processing, artificial intelligence, robotics, networked communication and cloud computing are transforming production and consumption, not least of knowledge itself, as well as the labour market. It is estimated that in the next 10–15 years, up to 50% of existing roles will be made redundant (Bridgstock 2017, 342). To take advantage of the new roles and opportunities that will be created, the university must radically transform, embracing internationalisation and transdisciplinarity. As Gerry Stoker and Mark Evans argue, the issue is not how academia draws up its dividing lines but rather which types of research can contribute to the problems we confront (2016, 2). Internationalisation provides essential skills in cross-cultural communication and promotes global citizenship. Transdisciplinary teaching offers a promising seed of knowledge-network activity. Teaching and learning methods that include student-centred, problem-based, practice-oriented and community-based learning (Stütz et al. 2014, 34) must become mainstream, while also including international partnerships and global outlooks and perspectives.[4] In this chapter I have discussed ways in which the internationalisation and the STEAM cultivation agendas overlap and how they can benefit from each other, suggesting practical ideas that can be implemented to this end. Internationalisation and transdisciplinarity are important ways forward to make the university future-capable.

Works Cited

Altbach PG, Yudkevich M (2017) International faculty in 21st century universities: themes and variations. In: Yudkevich M, Altbach PG, Rumbley LE (eds) International faculty in higher education: comparative perspectives on recruitment, integration and impact. Routledge, London, pp 1–14

Barry A, Born G (eds) (2013) Interdisciplinarity: reconfigurations of the social and natural sciences. Routledge, London

Beelen J, Jones E (2015) Redefining Internationalization at Home. In: Curaj A, Matei L, Pricopie R, Salmi J, Scott P (eds) The European higher education area. Springer, Cham, pp 59–72

Brewer E, Leask B (2012) Internationalization of the curriculum. In: Deardorff DK, de Wit H, Heyl JD, Adams T (eds) The Sage Handbook of International Education. Sage, London

Bridgstock R (2017) The university and the knowledge network: a new educational model for twenty-first century learning and employability. In: Tomlinson M, Holmes L (eds) Graduate employability in context. Palgrave, New York, pp 339–358

Brown VA, Harris JA (2014) The human capacity for transformational change: harnessing the collective mind. Routledge, London/New York

[4]An example of curriculum developed for STEAM and internationalisation is the University of Northern Arizona's Global Science and Engineering Program (GSEP), launched in 2011 to develop global capabilities. Students are required to complete a second major in an Asian or European language simultaneously and spend their fourth year abroad, half of it in a placement, and their fifth writing both capstone projects. Students indicated that their language major 'enabled them to use a different part of their brain' (Killick 2017, 200–201).

Brown VA, Deane PM, Harris JA, Russell JY (2010) Towards a just and sustainable future. In: Brown VA, Harris JA, Russell JY (eds) Tackling wicked problems through the transdisciplinary imagination. Routledge, London/New York, pp 3–15
Buechley L, Peppler K, Eisenberg M, Kafai Y (2013) Introduction. In: Buechley L, Peppler K, Eisenberg M, Kafai Y (eds) Textile messages: dispatches from the world of e-textiles and education. Peter Lang, Frankfurt, pp 1–5
Carroll J (2015) Tools for teaching in an educationally mobile world. Routledge, London
Denicolo P (2013) Developing transferable skills. Sage, London
Frodeman R, Thompson J, Mitcham C (eds) (2012) The Oxford handbook of interdisciplinarity. Oxford University Press, Oxford
Graff HJ (2015) Undisciplining knowledge: interdisciplinarity in the twentieth century. Johns Hopkins University Press, Baltimore
Green W, Whitsed C (2015) Introducing critical perspectives on internationalising the curriculum. In: Green W, Whitsed C (eds) Critical perspectives on internationalising the curriculum in the disciplines. Sense Publishers, Rotterdam, pp 3–22
Guldin R (2016) Translation as metaphor. Routledge, London
Herath D, Kroos C, Stelarc (eds) (2016) Robots and art: exploring an unlikely symbiosis. Springer, Cham
Katz P (ed) (2016) Drawing For science. Sense, Rotterdam
Killick D (2017) Internationalization and diversity in higher education: implications for teaching, learning and assessment. Palgrave, New York
Lawrence RF (2010) Beyond disciplinary confinement to imaginative transdisciplinarity. In: Brown VA, Harris JA, Russell JY (eds) Tackling wicked problems through transdisciplinary imagination. Routledge, London, pp 16–30
Leask B (2013) Internationalization of the curriculum and the disciplines: current perspectives and directions for the future. Journal of Studies in International Education 17(2):99–102
Leask B (2015) Internationalising the curriculum. Routledge, London
Leyton M (2006) The structure of paintings. Springer, Wien/New York
Stoker G, Evans M (2016) Introduction. In: Stoker G, Evans M (eds) Evidence-based Policy making in the social sciences: methods that matter. Policy Press, Bistol, pp 1–12
Stütz A, Green W, McAllister L, Eley D (2014) Preparing medical graduates for an interconnected world: current practices and future possibilities for internationalising the medical curriculum in different contexts. J Stud Int Edu *19*(I):28–45
Tomlinson M (2017) Introduction: graduate employability in context, charting a complex, contested and multi-faceted policy and research field. In: Tomlinson M, Holmes L (eds) Graduate employability in context: theory, research and debate. Palgrave, New York, pp 1–40
Trowler P (2012) Disciplines and interdisciplinarity: conceptual groundwork. In: Trowler P, Saunders M, Bamber V (eds) Tribes and territories in the 21st century: rethinking the significance of disciplines in higher education. Routledge, London, pp 5–19

Chapter 11
What Can We Learn About STEAM from Bridges?

David Blockley

Abstract Science is used, especially by the media, as an umbrella word, cardinal and all embracing. Technology is widely regarded as the hand-maiden of science – simply its application. Intellectuals have seen little of interest in the differences. Engineering is barely understood and its art and relationship with medicine goes unrecognised. Bridges are a half-way house between the visible simplicity of early tools and the 'black box' complexity of modern technology. Engineering is done by people for people.

Keywords Bridges · Art and science bridging · Art and engineering · Art and science · Complexity

Introduction

Technical and social issues are intimately linked. An example is that governments have to consult technical experts – scientists and engineers – to formulate policy about national issues such as flooding, transport and the security of cyberspace. Public information about such issues relies heavily on media reports and the manifestation of science and technology in almost all of the tools of modern life. From the food we eat, the water we drink, the transport we travel in to the medical care we get, science, technology, engineering and mathematics (STEM) are critical. Much of it is controversial for all sorts of reasons. Three examples will suffice to make the point. First, the performance of the railways is under constant criticism and scrutiny over public spending as lines are upgraded. Integrated transport planning is difficult. Second, our reactions to extreme weather events such as flooding and how much those events are influenced by climate change are understandable but solutions are not easy. Third, in the UK, the performance of the National Health Service is pressured not just by lack of public funding and the complexity of streamlining systems

D. Blockley (✉)
Emeritus Professor of Civil Engineering, University of Bristol, Bristol, UK
e-mail: D.Blockley@bristol.ac.uk

© Springer International Publishing AG, part of Springer Nature 2019
A. de la Garza, C. Travis (eds.), *The STEAM Revolution*,
https://doi.org/10.1007/978-3-319-89818-6_11

but also by increasing costs of technical treatments such as drugs or sophisticated equipment like scanners.

Over many years a number of surveys have indicated the difficulty of recruiting young people into the professions that develop and use scientific expertise. The reasons are complex. Partly as a result, at the end of the last century, those disciplines united under the banner STEM. They did this in order to present a common message about the attractions of a career in the related disciplines. Other letter combinations such as STM, eSTEM (environmental STEM) and STREM (with R for robotics) were suggested, but STEM emerged as the preferred option for most people. In this chapter I will argue that this unity has not been wholly positive particularly for engineering and that there is an urgent need for the letters to be unscrambled. I will also welcome the additional A for art in STEAM but suggest adding another M for medicine to make STEAMM. The chief reason is that STEM has come to be dominated by S – science. The word science is used, especially by the media, as an umbrella word, cardinal and all embracing. It smothers and obfuscates technology and engineering and is, in recent times with the big advances in the biological sciences, in danger of smothering medicine too. As we face the complex issues of the twenty-first century, this primacy of science is not helpful for reasons that need to be articulated.

One manifestation is that technology is widely regarded as the handmaiden of science – simply its application. Just one example is the landing of Philae on Comet 67P from the Rosetta probe in 2014 (Blockley et al. 2014). It was extensively reported as a scientific achievement. There was little or no recognition in the media of the engineering that made it possible. The reporting prompted myself and colleagues Professors Stuart Burgess and Paul Weaver of the University of Bristol, UK, to write to the *Guardian* newspaper. The letter was published with the headlines 'Let's hear it for the Rosetta engineers'. We said that the landing was a fantastic achievement, a tremendous scientific experiment based on wonderful engineering. Engineering is the turning of a dream into a reality. So please give credit where credit is due – to the engineers. The success of the science is yet to be determined, depending on what we find out about the comet. Engineering is not the handmaiden of physics any more than medicine is of biology – all are of equal importance to our futures.

One of the basic reasons for this disregard of engineering is that intellectuals have seen little of interest in the differences between science and technology. Engineering is barely understood and its art goes unrecognised. An article in the *New Scientist* is typical: It can be hard to convince outsiders that engineering is a worthwhile profession. Perhaps that's because the name 'engineer' has come to be attached with men (rarely women) in overalls, fixing phones or mending bits of machinery. In a similar but different way, mathematics has not fared well. In the media it is portrayed as hard and difficult and not something ordinary people need beyond everyday arithmetic. Publishers will advise that any mathematics in a popular science book will reduce sales – though there have been some notable exceptions. The writer Siri Hustvedt (2016) puts her finger on the pulse when she acutely observed:

> [...]in popular culture science is often perceived as monolithic [...] while everyone is aware that "scientists" come from many disciplines and frequently change their minds [...] there is a powerful sense in which that they, the scientists, are on an inexorable march forward [...] as it (science) methodically uncovers the secrets of nature.

It is commonplace for engineering and technological success to be quoted as the reason why science is true. In the minds of the many (and especially the media), the landing of Philae proves the truth of science. Of course as Siri Hustvedt (2016) says, the reality of the relationship between our scientific theories and the natural world is undeniable, but it has had a *blinding* effect on many. They fail to see that science rests on assumed foundations about material reality and most importantly is *incomplete*. That practical technical success proves the truth of science is a myth and wildly misleading. Engineers (and medics) have to go way beyond science to achieve their goals and in major instances such as the development of steam engines the science came after the engineering. Few people have taken on board the logical message of the philosopher Karl Popper (1934). One of his central tenets was that no matter how many observations concur with a theoretical prediction, we cannot conclude that the theory is true. We can conclude only that it is *highly tested*. One cannot rule out the possibility of a future test that will show the theory to be false. One test failure shows the theory to be false. In practice of course, it is rare for a theory to be straightforwardly falsified. Matters are rarely simply true or false despite the assumptions of deterministic science. When we succeed, we show how a theory works in a context and how it may well be dependable for use in that context. One rather obvious example is Newtonian mechanics. Whilst Einstein showed through relativity theory that Newtonian mechanics does produce incorrect results in certain circumstances (bodies travelling at a significant fraction of the speed of light), the designers of earth-bound structures such as bridges can depend on Newton's laws since bridges do not fall into that category. Nevertheless, there are other more detailed assumptions that have to be made by bridge designers within the science of structural mechanics. These may well be problematic and have practical consequences. In summary, the success of bridge designs around the world does not prove the truth of Newtonian mechanics. What they do demonstrate is the *dependability* of Newtonian mechanics *in a context.*

STEAM, with an emphasis on the A for art, reminds us that the professional activity of engineers and medics is not simple applied science. Rather it is creative doing based on informed knowing. Like medicine, law and economics, engineering is done by people, for people, by making changes to improve the human condition. But the objects, knowledge, activities and skills needed to satisfy humans include devices such as legal documents and economic regulations that are often complex instruments for making changes that require the attention of specialists. David Howarth, a distinguished academic lawyer and politician, has written about transactional and legislative law (Howarth 2013): 'like engineers [...] lawyers want to make something useful that works for their clients. We need to garner more support for STEAM by arguing that we need to reconnect across technical and social boundaries'. As we face the as-yet unknown consequences of the challenges of the twenty-first century, such as climate change, we have to collaborate more and learn together better. I am not suggesting technocracy – government by an elite of technical experts. Technocracy means that leaders are chosen for their expertise and unelected and it is undemocratic. We need to improve the use of technical expertise in democratic decision-making. When governments consult experts, there is often a lack of mutual understanding. For example, despite being advised to the contrary, governments in the UK did not seem to realise that lack of continuity of policy in providing and maintaining infrastructure created extra and avoidable costs.

I want engineers to play a full part in public decision-making, but to be effective I believe that there will have to be a change in the current culture of engineering education and professional development. There will need to be a much greater empathy, connection and integration with other disciplines, especially the arts, social sciences, law and medicine and with science and philosophy. Engineers will need to work much harder at explaining to lay people what they do and how they do it. They will also need to develop a better capacity to listen and learn from them in return. Together, we have to realise that engineering has been and will continue to be integral to the flourishing of humanity. Engineering education can be education for life as well as for a vocation. Teachers of engineering need to learn both to teach to engineering (i.e. as a vocation) and through engineering (as a general education).

The American philosopher Carl Mitcham (2014) has challenged engineers to develop what he calls engineering self-knowledge. He relates it to C. P. Snow's assertion about a tension across the yawning gap between the two cultures of science and the humanities. In his 1959 Rede Lecture at the University of Cambridge, Snow said that the split was a major hindrance to solving the world's problems. Mitcham's gap is not between two forms of knowledge production but rather between two forms of practical action and consequences. He contrasts designing and constructing the world, by which he means engineering, with reflecting on what it means – by which he means the humanities. At root Mitcham wants a much wider debate across the gap between engineering work as performed in changing our physical world and the broad social context with which it interacts.

As I have said, engineering and medicine are not the handmaidens of science. Science puts the stress on – defines its purpose as – context-free knowing. In my view that is not possible – science is inevitably dependent on context. Unfortunately, the widely held common view of popular culture rarely concurs. In a similar misunderstanding, engineering stresses – has the purpose of – doing and so is widely held in popular culture to be manual. The history of how we humans have flourished shows this is wrong and the separation of knowing and doing is harming us. In the light of modern understanding of what science can and cannot deliver, we need to return to the ancient view that knowing and doing are intimately related – they leapfrog over each other as we progress. We need to appreciate *how* the history of the divergence of the professions has led us to some deep confusions. For example, when people who say they are scientists make things – they are actually performing feats of engineering although they call it science. Similarly, when people who say they are engineers discover new knowledge – they are actually doing science (mainly academics) although they call it engineering.

Engineering is both a people profession and an art in the sense that it speaks to our emotional needs, good and bad. Engineering, like medicine, is change *making* – producing something that didn't exist before – such as a bridge or a new treatment. In preparation for the next section, we should note that the Greek word for making is *poiesis* and the root of the word poetry. In brief *poiesis* is about reconciling thought with being – creating something wonderful where before there was nothing. The purpose of the making is a 'bringing forth,' i.e. the making is the means to a particular, valued end—a poem as an aid to understanding, a bridge as an aid to a crossing or a heart pacemaker as an aid to save lives. Poets can be engineers and medics and vice versa.

A Little History

The ancient Greeks had some key insights which over the centuries have either been lost or misinterpreted. Some of them can help us to disentangle STEAM. I focus here on Aristotle, his notion of practical wisdom and its relationship with art. *Aristotle defined two kinds of virtues*. First are the intellectual virtues such as wisdom, understanding and practical wisdom. Second are the moral virtues such as liberality or beneficence and temperance or self-control. He described virtues as dispositions or attitudes. Disposition comes from the way we behave – including habits and the way we deal with situations. Aristotle wrote 'For example, the excellence of the eye makes both the eye and its function good (because it is through the excellence of the eye that we see well) [...] human excellence will be the disposition that makes one a good man and causes him to perform his function well' (Thomson 1955). Aristotle's word for the virtue of practical wisdom was *phronesis*, a concept that we seem to have lost.

Aristotle saw five ways of arriving at the truth. They were art (*techne*), science (*episteme*), intuition (*nous*), theoretical wisdom (*sophia*) and practical wisdom (*phronesis*). *Techne* was about production but not action. Nowadays we tend to see production as part of action. But *techne* has given us the modern words, technical, technique and technology that imply mere technique or rule-following. Technical in this modern sense implies manual skill often devoid of creativity – but that is not what Aristotle meant. Science or *episteme* has given us the word epistemology or knowledge. To Aristotle *episteme* was the very powerful idea of knowledge that cannot be otherwise than it is – eternal and universal truth that cannot come into being or cease to be. Modern physics has demonstrated that this view is no longer tenable. Quantum mechanics has revealed deep uncertainties in what we can know. Intuition or *nous* was a state of mind, similar to our modern notion of mind, intelligence, intellect or even common sense. Perhaps most importantly for our purpose here, Aristotle distinguished between theoretical wisdom as *sophia* and practical wisdom as *phronesis*.

Aristotle proposed three *categories of human activity*. He called them *praxis*, *poiesis* and *theoria*. *Praxis* means practice – as distinguished from theory. *Poiesis* means making or forming and reconciling human thought and being with the world. *Theoria* means contemplation and the root of the word theory. The intellectual faculties required were *phronesis* for *praxis*, *techne* for *poiesis* and *sophia* and *nous* for *theoria*.

It is important to understand that in the minds of the ancient Greeks, *theoria* had total priority because *sophia* and *nous* were considered to be universal and eternal. Everything else is variable, finite, contingent, uncertain and therefore inferior. As I have said, modern science requires us to think about science as *sophia* quite differently because developments in quantum mechanics demonstrate that we cannot escape uncertainty. It is now actually a matter of faith as to whether we think science as eternal, universal, value and context-free *sophia*. By faith I mean belief not based on proof. Most modern scientists emphasise the importance of the scientific method rather than the universality of the results they find. We now see scientific theories as incomplete theoretical models. They work in a context that has to be understood. They are not true in all possible contexts, because there will be situations in which the models have not been tested because we haven't identified them yet – the unknown unknowns.

Praxis included culturally shared knowledge but required *phronesis*. *Poiesis* was the act of making or producing something. Martin Heidegger in his 1977 essay 'The question concerning technology' (Heidegger 1977) was influential when he controversially argued for the 'primacy of practice' over science. He was later to describe *poiesis* as 'a bringing forth' or 'a revealing', something done for a purpose where the making is not an end in and of itself. Rather it is a means to a particular end – for example, a pair of shoes is an aid to a good life for Peter because Peter needs shoes.

What did Aristotle actually mean when he referred to *phronesis*? As I see it, *phronesis* is a means towards an end arrived at through moral virtue. It is concerned with 'the capacity for determining what is good for both the individual and the community'. It is a virtue and a competence, an ability to deliberate rightly about what is good in general, about discerning and judging what is true and right, but it excludes specific competences (like deliberating about how to build a bridge or how to make a person healthy). It is purposeful, contextual but not rule-following. It is not routine or even well-trained behaviour but rather intentional conduct based on tacit knowledge and experience, using longer time horizons than usual and considering more aspects, more ways of knowing, more viewpoints, coupled with an ability to generalise beyond narrow subject areas. *Phronesis* was not considered a science by Aristotle because it is variable and context dependent. It was not an art because it is about action and generically different from production. Art is production that aims at an end other than itself. Action is a continuous process of doing well and an end in itself in so far as being well done it contributes to the good life. I see therefore that *phronesis* is about being a good professional. Professionals don't profess to know it all. Rather they have learned lessons by reflecting on extensive experience of rational life, with all of its risks and uncertainties. These include unintended consequences as well as the rich lightness and darkness of our emotional experience. Cleverness is not sufficient for *phronesis* – it requires life's experience both rational and emotional.

Before the Renaissance Art, Craft, Engineering and Architecture Were One

The separation between art and engineering came only after the Renaissance as our scientific knowledge grew more powerful. Up to this time, the only theoretical language for practical work was geometry. But progress in science during the Renaissance was soon to change that. There were several consequences. First the Greek idea that science is 'the Truth', i.e. eternal and unchanging prevailed, and so science was regularly in conflict with religious faith. Second, in Renaissance Italy, patrons who wanted monumental buildings began to recognize that artisans, who had always worked under the direction of guilds or the church, were not only skilled technicians but also creative thinkers, discoverers and inventors. Third, from about 1700–1900, the processes of separation between architecture and engineering that started with the artist engineers of the Italian Renaissance gained momentum. Specialisms emerged to meet new needs. Two different points of view began to appear – one emphasising aesthetic appearance and the other the structure. Fourth, we have struggled with our understanding of what makes creative art. It is complex and a matter of constant discussion. We usually know what we like when we see it but are unsure of the value of the art we do not like. Art illuminates us and enriches our emotional experiences. Art impacts on our social wellbeing, cohesion, our health, our educational system and our economy. Art comes in many different forms and structures of matter such as paint on canvas, carving of wood or stone or even a large metal structure like Antony Gormley's 'Angel of the North'. Fifth, engineering has come to be regarded as applied science – technology. Philosophers have found no interesting differences between science and technology (interpreted purely as applied science), and so intellectuals and the media have subsumed it under science and called it technology. Engineering was downgraded in popular culture to craft – but not craft in its creative sense but more akin to vocational technique and manual skills devoid of creativity. This has been and still is a major category mistake.

What Can We Learn from Bridges About STEAM?

Bridges are interesting for many reasons. One is that they are a kind of half-way house between the visible simplicity of early tools (like the plough) and the hidden 'black box' complexity of modern technology (like computers and jet engines). Older tools are so simple that the way they function is obvious to lay people without technical training, for example, a hammer, a saw and a plough. Their functioning contrasts strongly with a computer or smart phone tools with interior operations that are opaque and mysterious to anyone without special knowledge. Bridges are somewhere in between – at once simple like a plank across a stream but also complex like a cable-stayed bridge with intuitive flows of force but actually massively redundant, i.e. not easily calculable.

What we lack in modern public conversation is a means by which interested people can have a conversation about the functioning of the complex objects we all rely on. Furthermore, this extends to the complex systems critical to climate change such as flooding and extreme weather like wind. We need to provide the basis for a conversation based on some level of understanding of how things work and hence a more informed understanding of what is possible to change them – especially when resources have to be prioritised.

I think that one of the candidates for the language of that conversation is energy. In all systems, there is a potential that drives flow against an opposition. In this chapter I will describe these processes for bridges. I will try to show how forces flow through a structure. It will hopefully illustrate the art of STEAM. Hopefully that will also stimulate further work on a greater overview understanding how modern equipment such as smart phones work and perhaps more importantly how complex river systems can be managed to minimise the effects of flooding (Blockley 2017) and stimulate other similarly important inquiries.

Making the As/Is Distinction

The drawing of boundaries around systems is not straightforward. Our scientific theories are models of reality, not the reality itself. There are still considerable gaps between what we know, what we do and why things go wrong. Therein lies the creative art – original synthetic problem solving and analytic judgement through a legal duty of professional care.

In 1739 David Hume, the Scottish philosopher, made a famous distinction between is and ought. He pointed out that many people make false claims about what ought to be the case, on the basis of what *is* the case. In practical applications of science, there is often a similar but largely unrecognised distinction – I call it the *as/is* distinction. It is the difference between a theory or model being used *as if it is true* and whether a theory *actually is true*. When we act in a professionally considered way, we do it based on a conscious decision. If that act is successful – in other words things turn out as we expect – all well and good but what can we conclude about the verity of the decision? If the decision was made based on a theory used as if it is true, we cannot conclude that in reality it is actually true – even though the act was successful. The key to appreciating the *as/is* distinction is context. If we are confident we understand the context sufficiently well, then we can make a decision as if the theory on which it is based is actually true. If we stray outside of that context, we must be careful because we will not be acting with a duty of care. For assumptions about Newton's laws applied to bridge building, there is no concern, but for many others, the situation is much less straightforward – there are assumptions that do matter. For example, the steel of which many bridges are built is assumed to have a linear relationship between the force applied and the resulting amount of stretch. But as the force gets bigger, the graph becomes non-linear. Many other common materials like concrete and aluminium do not have anything like a linear relationship, yet many successful bridges have been designed assuming those materials are linear.

Practical Rigour

The scientific method is rigorous. Practice is often criticised as being ad hoc and lacking rigour. Engineers are often criticised as non-rigorous because they rely on approximations and judgments. Engineers are actually rigorous in a different practical way, for two impelling reasons. Firstly, engineering products will inevitably be subject to the ultimate test – mother-nature. If a structure is inadequate, it will collapse. This imposes the kind of 'natural honesty' since it is a requirement that cannot be twisted by propaganda or 'spin'. Secondly, engineers have a legal duty of care to society. Under this duty engineers have to justify their decisions in public if called upon to do so.

Rigour is the strict enforcement of rules to an end. Mathematical logic is the ultimate form of absolute rigour: it has one value – truth. It is top-down reasoning, i.e. theorems are deduced from axioms (including rules) which are true by definition. The result is a self-consistent body of true statements – but only if the axioms correspond to reality. Science is bottom-up reasoning. We predict some observable phenomena and test it in an experiment. If the theories we use are formulated by generalising other observations, we call it induction. However, we generate our theories the emphasis in physical science is on testing to find observable, precise truth. But since truth is correspondence to the facts, we get an infinite regress – facts are true statements and we are defining something in terms of itself.

Practical rigour is the meeting of a need by the setting of clear objectives. I see seven elements in practical rigour. First comes the need to make something work. Creating practical solutions to meet explicit needs. Delivering a system valued in a variety of ways, not just efficiency or cost but including aesthetics, sustainability, practicability and resilience. Second, it is creating appropriate theoretical and practical models through making sensible approximations that respect nature. Being aware of the *as/is* distinction. Leaving no stone unturned with no sloppy or slipshod thinking, showing a duty of care. Third, the whole as well as the parts has to be considered. The scientific approach is a process of selective inattention. We break a problem into its separate components, take out the difficult bits that we do not know how to solve and focus on what we can solve. Practical rigour requires us to deal with the bits of the problem that we do not always understand too well. Fourth is the making of judgments. Professional opinions are not arbitrary, they are based on dependable evidence. Fifth is a developing and evaluating of dependable evidence. The only clear way to judge the dependability of evidence is to subject it to as many tests as seems appropriate. Sixth is exercising creative foresight. Practice requires the creativity to imagine what might happen. Seventh is feedback and learning to improve. Practical rigour implies practical intelligence – which in turn implies practical experience. Experience is necessary but not sufficient for practical intelligence. And practical intelligence is necessary but not sufficient for practical rigour.

Engineers Create Technology by Configuring Patterns of Flowing Energy to Achieve a Human Purpose

Function and purpose are potential that creates a flow of change. Potential simply means capable of becoming for a reason, aim or intention. Potential in electricity is voltage and in mechanics is velocity. Flow in electricity is current (amps) and is force in mechanics. Of course, there is always some opposition to any flow, and some of the energy is dissipated or becomes irretrievable and no longer available to do useful work. There are three kinds of opposition to flow – collectively called impedance. The first is opposition through storage or accumulation of potential – called capacitance. Examples are an electrical battery which stores electrical charge, a water tank which stores a volume of water and the mass of an object which stores inertia. The second is delay through storage of flow – called inductance. Examples are an electrical coil such as the wire windings of an electrical motor, a waterwheel and the flexibility of a structure. The third is frictional resistance or dissipation of energy and any energy that is no longer available to do any work – the technical name for that is entropy. Examples of energy dissipation are electrical resistance which, for example, in a bar heater emits radiant heat, mechanical friction and damping as in a car shock absorber or a sand filter in the flow of water.

How do engineers configure flows of energy to achieve a purpose? They do it by understanding how a potential creates a flow and they use various forms of opposition to flow to create the effects they want. A bridge is rather like a pile of books on a shelf – a pile of stored potential energy. The energy only becomes apparent if the bridge collapses. Within the bridge the forces of the potential energy flow through the structure rather like water flows through pipes until it is transferred to the ground. Structural engineers control that potential energy (which engineers call strain energy) by configuring structural elements that oppose the flow as required. The main parameters to do this are mass, stiffness and damping.

To make a bridge stand up safely, the first requirement is that whatever mother-nature throws at the structure, it needs to stand firm. The demands come from wind, earthquakes, the weight and activities of people, road traffic, trains or crashes such as a railway accident, ship docking out of control or the action of huge sea or storms or any other natural event such as lightning. Demands may come from the legitimate activity of people like crossing a bridge with a weight limit or illegitimate activity like excessive speeding. So the engineer has to estimate the magnitude of these demands and how that demand relates to the strength of the structure. But how far should it go – does it include tsunamis and terrorist attacks such as 9/11? The answer is that the engineer must estimate what it is reasonable to assume. The engineer must design and build the bridge so that the resistance to those demands is greater than they are by a suitable margin of safety. The chances of collapse must be acceptably low – but they are never zero.

Let's illustrate these points simply by looking at the simplest of examples – a plumb bob hanging on a string or a tug-of-war competition or a cable in a hanging bridge. The string, rope and wire are being pulled by the weight of the plumb bob, the opposing pulls of the tug-of-war teams or the weight and traffic loads on the bridge. Length has no effect on the tensile strengths of the materials. Their strength depends on two main factors – the material of which it is made and its diameter. The stronger the material (steel wire is stronger than string) and the larger the diameter, the heavier the plumb bob that the string can withstand. Likewise, and rather obviously, the rope for a 12-a-side tug-of-war will have to have a much larger diameter than the string for a plumb bob.

What is going on inside the string, wire and rope? Let's do a thought experiment. Let's imagine that you are holding a string with nothing at the other end. You pull on it and the string will simply move. Now imagine someone else (A) pulling back on the other end of the string – the string becomes taut. The string is transmitting two forces – your pull which we call an action and A's pull – which we call a reaction. If your action force and A's reaction force are equal, then the string does not move – action and reaction are equal and opposite – balance and equilibrium. How strong does the string have to be? A further thought experiment will help us sort it out. Imagine what would happen if you cut the taut string with scissors. Both cut ends would become free and the strings would move. Now imagine asking someone else (B) to take the place of the cut fibres of the string by pulling on each end at the same time. Now we have two pieces of string both balanced in equilibrium. The force that B is now providing is called an internal force. This force balances the external force imposed by you at one end and A at the other.

How do we interpret this more generally? When external forces are applied to a structure (action – you pulling the string), they will be resisted (reaction – the pull at the other end), and connecting these two groups of forces are sets of internal forces as shown in Fig. 11.1a. All of the forces are in balance or equilibrium when nothing moves. We can develop that idea and think of Fig. 11.1 in a different way. Imagine we had cut the string in two places so that the string between the cuts represents not the total length of string but just a small piece of it. The diagram (Fig. 11.1b) looks exactly the same but the forces labelled as external would now be internal and the diagram would now show balanced internal forces. In effect, we are

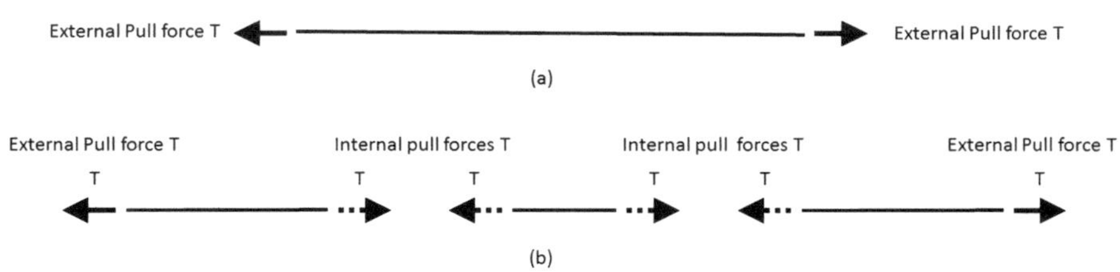

Fig. 11.1 The internal forces in a string pulled in tension

showing internal forces flowing along the string. In the simple example of tension, the internal forces do not change along the length of the string. We find that in more complicated cases, they do change. In that case, we usually examine the effect on a very small area of material such as a square millimetre and we call that a stress. The idea of cutting a piece of the structure out and finding the internal forces on a small section is an important one for structural engineers. It is why engineers refer to forces flowing through a structure rather like water through a pipe or electric current along a wire.

Just as a piece of string will snap if you pull too hard on it, then any part of a bridge may break if the internal force effect gets to be too large. In general, there are three ways in which materials are strong in different combinations – pulling, pushing and sliding or tension, compression and shear. For more details on this for the general reader, see Blockley (2010).

The engineer has to ensure that three independent conditions are satisfied. First all the forces acting horizontally must balance; second the vertical forces must balance; and third the turning forces must balance. Indeed these are the three independent conditions to establish equilibrium in two dimensions (we haven't considered three dimensions for simplicity – the principles are the same but the execution is more complicated). If we balance the forces in any other direction, we will not get any more information since the equilibrium equations will not be independent. There is another complication however. Sometimes we cannot find the internal forces using equilibrium only. It turns out that equilibrium between internal and external forces is necessary, but it may not be sufficient if we want to find all of the internal forces. Three further conditions apply. First, as we have already seen, we must have equilibrium. Second we need to know how the material responds to the internal forces, for example, how it stretches under tension – these are called constitutive relations. Third we must make sure that the various bits of the structure remain properly connected, i.e. all displacements are compatible. To incorporate these extra requirements, we need now to look at equilibrium slightly differently. When a structure reacts to a load, every part of it, no matter how little or large, moves or deforms very slightly and in doing so stores potential energy. Potential energy is due to position and is created when something is displaced. You can witness the release of this potential energy when a structure breaks. Just imagine what would happen if you were pulling on a tight string or heaving in a tug-of-war and it snaps – you will fall backwards as the energy is released. This locked-in potential energy is called strain energy. It turns out that when a structure is in equilibrium, the strain energy is at a minimum.

It is clear, I hope, from this illustration that the whole process is much more than an application of known science to a well-understood situation. Judgements have to be made. Despite its apparent complexity, the scientific model is an approximation to reality. The gaps between the theory (what we know), the practice (what we do) and why the bridge might fail are not insignificant. They are filled by the A of art in STEAM. Engineering is an art that uses science to do things for people. Bridges are built by people for people and are of great significance to the functioning of society.

Conclusions

1. A shortage of recruits of young people into engineering is endangering our collective futures as we face up to the challenges of more extreme events due to our changing climate.
2. The unity of STEM has not been wholly positive particularly for engineering. The main reason is that it is dominated by S for science. Science is used, especially by the media, as an umbrella word, cardinal and all embracing. It smothers and obfuscates technology and engineering. I welcome the additional A for art in STEAM but suggest adding another M for medicine to make STEAMM.
3. There is an inevitable gap between theory and practice – between what we know what we do and why things go wrong. It is filled by the A for art. Aristotle's understanding of phronesis as practical wisdom helps us to see that theory and practice advance hand in hand.
4. We need to rebrand engineering as an activity (like medicine) which is done by people for people to improve the human condition.
5. We lack public conversation about the functioning of the complex objects we all rely on. I think that the basis for the language of that conversation is energy. Potential creates flow against an opposition that stores potential, stores flow and dissipates energy.
6. Bridges are a kind of half-way house between the simplicity of early tools like the plough and the opaque complexity of modern tools such as computers. Through them we may be able to initiate a public conversation that may help lay people better appreciate the complex things that they rely on.
7. By appreciating better how energy flows in bridges, we can perhaps begin to see the crucial role of human purpose and the embedment of models in physical, human and social systems. This may also help lay people to understand better the relationship between the different components of STEAM (M).

References

Blockley DI (2010) Bridges: the science and art of the world's most inspiring structures. Oxford University Press, Oxford

Blockley DI (2017) Turning dreams into reality. Oxford University Press, Oxford

Blockley D, Burgess S, Weaver P (2014) Let's hear it for the Rosetta engineers, Guardian newspaper 18th Nov 2014. See https://www.theguardian.com/theguardian/2014/nov/18/rosetta-all-blacks-hungary-gin-blackberries

Heidegger M (1977) The question concerning technology, harper perennial; Reissue edition 2013, London

Howarth D (2013) Law as engineering. Edward Elgar, Cheltenham

Hustvedt S (2016) A woman looking at men looking at women. Simon & Schuster, New York

Mitcham C (2014) The true grand challenge for engineering: self-knowledge, Issues in Science and Technology, vol XXXI Issue 1 See http://issues.org/31-1/perspectives-the-true-grand-challenge-for-engineering-self-knowledge

Popper K (1934) The logic of scientific discovery. Routledge, London

Thomson JAK (trans) (1955) The ethics of Aristotle, The Nicomachean Ethics. Penguin Book

Part IV
STEAM Teams

Chapter 12
STEAM at Work: Physiological and Psychological Perceptions of Risk of Cyclists

Ronan Doorley, Vikram Pakrashi, Eoin Byrne, Samuel Comerford, John A. Groeger, and Bidisha Ghosh

Abstract This chapter is concerned with a research project in which collaboration between psychology, medicine and engineering was necessary to assess the physiological and the psychological perceptions of risk in cyclists, comparing both to determine their impact on wearable technology that measures bodily signals of stress. Here we describe the research design and process, highlighting how insights from all these disciplines were necessary to carry out the project and interpret the conclusions.

Keywords Cycling · Physiological response · Psychological response · Wearable technology · Ireland · Heart rate · Risk perception · Travel diary · Transport · Infrastructure

---

R. Doorley · B. Ghosh
Department of Civil, Structural and Environmental Engineering, Trinity College Dublin, Dublin, Ireland

V. Pakrashi (✉)
Dynamical Systems and Risk Laboratory, School of Mechanical and Materials Engineering, University College Dublin, Dublin, Ireland
e-mail: vikram.pakrashi@ucd.ie

E. Byrne · S. Comerford
Department of Civil and Environmental Engineering, University College Cork, Cork, Ireland

J. A. Groeger
School of Social Sciences, Nottingham Trent University, Nottingham, UK

© Springer International Publishing AG, part of Springer Nature 2019

A. de la Garza, C. Travis (eds.), *The STEAM Revolution*,
https://doi.org/10.1007/978-3-319-89818-6_12

Globally, cycling is being promoted as a sustainable mode of travel which has multitude of individual and societal benefits. The World Health Organization (WHO) has recognised these benefits and recommends cycling as an activity to improve both physical and mental health. Our study investigated this claim with the aim of reaching a better understanding of the relationship between physiological and psychological effects of cycling. At the intersection of engineering and psychology, this study explores whether the risk perceptions of cyclists are reflected in their physiology.

Cyclists completed a test route under normal traffic conditions in Cork, Ireland, and their heart rates and self-reported risk ratings were recorded in real time. We found that situations perceived as high risk are likely to elicit higher heart rates than situations which are perceived as low risk. The cyclists also kept travel diaries during 1 week of their normal commute, which recorded their risk perceptions in relation to the road elements and any specific incidents they encountered.

Introduction

Incorporating cycling into daily routine is increasingly being recognised as an effective means of improving physical activity levels (Cavill et al. 2008; Kahlmeier et al. 2010; Rutter et al. 2013). Regular physical activity is a major contributor to physical and mental health, and the promotion of active travel has been identified by WHO Regional Office for Europe as a supporting intervention for the prevention and control of non-communicable diseases (WHO 2011). However, in order to ensure that any programme of increased physical activity brings health benefits, both the quantity and quality of exercise taken should be considered. The American College of Sports Medicine (ACSM) has published specific recommendations regarding both intensity and total energy expenditure of exercise for maintaining fitness (Garber et al. 2011). Since intensity cannot be easily measured directly, the ACSM has also identified optimum heart rate (HR) zones for various levels of exercise intensity (Garber et al. 2011) based on a linear relationship between (HR) and oxygen consumption (VO_2) (ACSM 2011; Åstrand 1976; Åstrand et al. 2003; Bassett et al. 2012). A leading HR monitor manufacturer has also published cycling-specific HR zones (Polar 2013).

There are many commercially available HR monitors that are used during cycling and other exercise routines in order to target specific HR zones and calculate activity energy expenditure (AEE). However, HR may be elevated by emotional and environmental factors, which could produce non-linearities in the relationship between HR and oxygen consumption (VO_2) (Åstrand et al. 2003; Crouter et al. 2008). In 1992 one research (Levenson 1992) studied the differences in autonomic nervous system (ANS) responses between different emotions. Large increases in HR were consistently found in response to a variety of fear inducing stimuli. Cacioppo (2000) later performed a meta-analysis across 22 studies on the question of whether there are emotion-specific physiological patterns. It was found that fear was associated with higher HR responses when compared to happiness, sadness or disgust. Levenson (1992) suggests that the association of fear with increased HR may reflect a close association of fear with the motor program of 'flight'. The 'flight response, described in detail by Cannon (1929), refers to the fear-induced bodily responses which prepare an organism for the great exertions which may follow. A more recent study of drivers' emotions found that anxiety, but not anger or happiness, was associated with increased perceptions of risk and with increased HR (Mesken et al. 2007). Another recent research (Lerner et al. 2005) studied differences in HR responses to anger and fear in participants performing a difficult arithmetic task. Fear was positively correlated with HR, whereas anger was negatively correlated with HR. All this evidence indicates that if an activity arouses feelings of fear or anxiety, these emotions may have an increasing effect on HR.

Several recent studies have suggested that cycling is perceived by transport users as being an unsafe mode of transport, particularly in an urban commuting environment where cyclists are exposed to motor vehicle traffic (Lawson et al. 2013; Winters et al. 2011, 2012). The perceptions of safety by cyclists are influenced by a wide range of factors such as age, regularity of cycling, road type and attitudes of vehicle drivers (Lawson et al. 2013). Individual events such as conflicts with other road users and near misses are likely to be perceived as particularly high in risk, but there is little information available in relation to these types of events. This implies that cyclists are likely to experience varying levels of fear and anxiety due to their perceptions of risk while cycling in a mixed mode network and it can reasonably be expected that this fear and anxiety may lead to variations in their HR. Also, there is no evidence to suggest that any currently available HR-based exercise intensity or AEE measurement devices are capable of compensating for the effects of significant levels of fear and anxiety. Therefore, the perceptions of risk among cyclists in a mixed, urban environment may cause miscalculation of exercise intensity and AEE as measured by the aforementioned devices.

In this study, we investigated the relationship between HR and perceptions of risk among cyclists. Evidence of a link was demonstrated, showing that the response of cyclists to situations which are perceived as being high in risk is not just psychological but physiological also. The link between risk perceptions and HR may be an indication of the 'flight' response which is typically accompanied by other physiological responses such as a release of adrenaline. The link between risk perceptions and HR also raises questions regarding the accuracy of devices which rely on HR for evaluation of the benefits of exercise in the context of active travel. This is a clear example of a STEAM approach that considers perspectives from science as well as the humanities and incorporates teams from both for collaboration. In this chapter, we also aim to address some limitations of previous research into the relationship between risk perceptions and the types of road elements encountered and incidents which occur. Recent evidence has shown that there is disagreement between cyclists' perception of the safety of particular road elements and actual safety and that perceptions of safety have more influence on cycling modal share than actual safety (Dill and Voros 2007; Keegan and Galbraith 2005; Noland 1995; Parkin et al. 2007; Winters et al. 2012).

Previous researches into perceptions of risk among cyclists have used methods such as site interviews (Moller and Hels 2008), video clips (Klobucar and Fricker 2007; Parkin et al. 2007), test courses (Landis et al. 1997, 2003), surveys (Lawson et al. 2013; Leden et al. 2000) and simulations (Hughes and Harkey 1997). With the exception of Lawson et al. (Lawson et al. 2013), the tests mentioned above focus on specific sites (with which the participant may or may not be familiar) and consider a small number of variables. They are also conducted over short time frames, and so the likelihood of a particularly high-risk event such as a near miss occurring is low. One case-crossover study addressed these limitations by recruiting cyclists who had sustained an injury while cycling, but since participants were recruited based on hospital records, minor injuries and near misses were not captured (Winters et al. 2012).

In this chapter, we explore the risk perceptions of each individual cyclist in relation to their regular route. Each route is divided into discrete road elements so that the cumulative experiences of the cyclists with respect to each type of road element can be studied. Another key contribution of this study is the ability to capture and characterise any specific event which may be perceived by a cyclist as being high in risk, such as a conflict or near miss. The STEAM framework advocated in this volume is particularly suitable for such assessments.

In the next section of this chapter, we describe the collaborative approach taken in this research. We then describe the methodology and results of a partially controlled field study we carried out in Cork city. We then present the methodology and results of an uncontrolled study in which we asked participants to keep travel diaries about their regular commutes by cycling. The results of both studies are then discussed together, and the chapter is concluded with some final remarks on the implications of this work.

Collaboration

The research described in this chapter lies at the intersection of two very distinct fields of study: transportation engineering and psychology. Designing urban form to ensure the safety of road users generally falls under the responsibility of engineers as well as urban planners. Engineers are trained to deal with objective truths such as those provided by collision statistics. However, subjective perceptions of risk, even if they are inaccurate, can strongly influence travel decisions and therefore influence safety objectively. For these reasons, it is essential for engineers and urban planners to give serious consideration to the subjective risk perceptions of road users. This need led to the collaboration between researchers in engineering and psychology described in this study. For this work, the behavioural aspects and the expected changes, along with interpretation and controls around testing, were more accessible by the psychologists, while the instrumentation and some of the analysis were specific to the engineers. It was interesting to observe in this project that the common goal of understanding the problem, its interpretation and nuances, along with the comfort level in statistical tools for analysis, were important in terms of creating a common vocabulary for a successful collaboration (a point that is frequently mentioned in this volume). The difference in training was compensated for by a clear understanding of what was there to achieve and in understanding where the strength of each contributor lay in the process of arriving at the results. This was ensured through multiple, well-organised meetings towards understanding each other. While training in statistics created the basis for speaking in a common language, studies around methodological aspects of psychology were novel to the engineers, and discussions were dedicated towards understanding the core fundamentals. A similar discussion was around the analytical tools for engineers, but previous experience of psychologists in instrumentation of individuals was a major advantage in the collaboration. While some time was dedicated around connecting the teams to each other's discipline, previous experience of key researchers in the team in working in an interdisciplinary atmosphere helped the process significantly. The key challenge lays in clear delineation of expected work from each individual and the connections of such work that led to a well-understood destination in terms of a well-defined problem that was being looked at. For this work, the methodological differences were more than the conceptual or intellectual differences. For an elaboration on collaborative problem framing and consensus, see 'Managing Consensus in Inter and Transdisciplinary Teams: Tasks and Expertise' (Defila and Di Giuliu 2017, 332).

Partially Controlled Experiment

Experiment Design

For the partially controlled cycling experiment, we recruited 13 volunteers from University College Cork using noticeboard advertisements and the authors' networks Doorley et al. (2015). All 13 volunteers were 23-year-old males. This homogeneity of age and gender in the study group was intentional in order to limit the impact of differences in certain personal characteristics on the results, albeit at some cost in terms of generalisability. Each participant completed a short questionnaire, the results of which are outlined in Table 12.1. As per the responses to the questionnaire, the participants were of varying degrees of physical fitness and cycling experience.

We asked the participants to cycle two fixed routes in Cork city while exposed to normal traffic conditions. Route 1 (anticlockwise) is shown in Fig. 12.1, and route 2 was a clockwise version of the same journey, beginning and ending at the same point. The lengths, road classes and annual average daily traffic (AADT) of each road section encountered on the routes are shown in Table 12.2. AADTs were calculated from short-term traffic counts using the NRA Permanent Counter Method (National Roads Authority 2012a) where possible and the NRA Generic Expansion Factor Method (National Roads Authority 2012b) otherwise. No traffic count data were available for one of the road sections, Gaol Walk. Each route was approximately 2.7 km long and took between 9 and 15 min to complete depending mainly on traffic signalling. Twelve scheduled rating points were chosen on each route to

Table 12.1 Results of questionnaire for participants of partially controlled experiment

Participant	Height (m)	Weight (kg)	BMI (kg/m²)	Physical fitness rating (1–10)	How often do you cycle?[a]	How long have you been cycling for? (years)
A.W.	1.85	79	23.08	6	3	16
B.R.	1.93	88	23.62	8	1	NA
C.G.	1.78	86	27.14	6	2	17
C.L.	1.91	86	23.57	10	1	NA
D.OD.	1.83	92.08	27.5	6	1	NA
E.B.	1.83	86	25.68	9	3	17
E.McN.	1.88	114.3	32.34	3	1	NA
M.P.	1.8	82	25.31	9	1	NA
O.K.	1.8	82.5	25.46	4	4	4
R.C.	1.8	75	23.15	5	2	12
R.McI.	1.8	73	22.53	5	2	1
S.C.	1.88	82.6	23.37	5	2	11
S.P.	1.78	80	25.25	7	1	NA

[a]Options were 1 (never), 2 (a few days a week), 3 (5 days a week) or 4 (everyday)

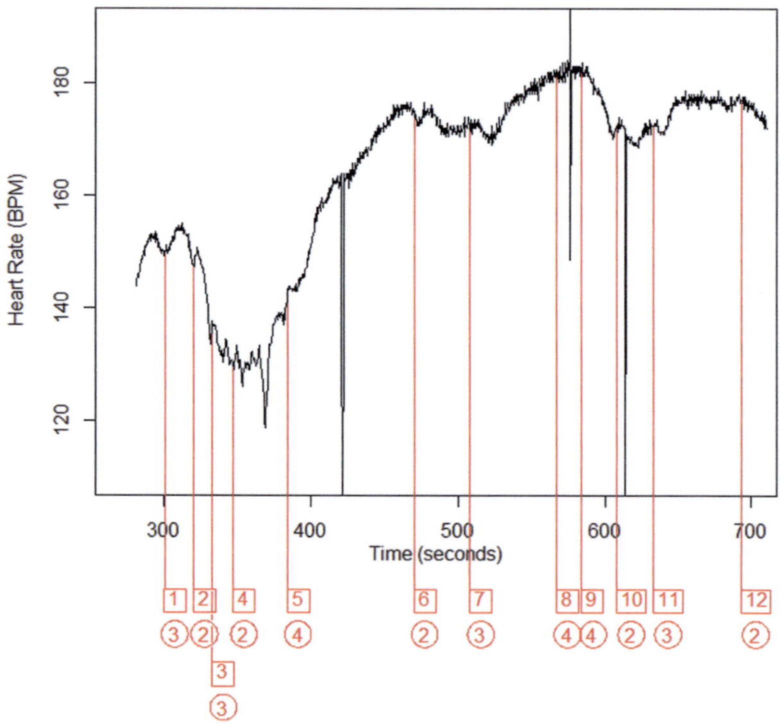

Fig. 12.1 Heart rate trend and risk ratings for subject O.K. while completing route 1 of the partially controlled experiment. Rating point identifiers are in squares, and risk ratings are in circles

Table 12.2 Roads encountered on routes in partially controlled experiment

Road	Class	Length (km)	AADT
College Road	Local road	0.75	10,251
Magazine Road	Regional road	0.3	11,595
Victoria Cross Road	National secondary road	0.45	19,762
Carrigrohane Road	National primary road	0.17	26,445
Western Road	National primary road	0.55	17,840
Gaol Walk	Local road	0.35	N/A[a]

[a]No traffic count data were available for Gaol Walk

include a variety of road elements (intersections, T-junctions, roundabouts, advanced stop lines, bus lanes, narrow roads, poor road surfaces) and manoeuvres (right/left turns, intersection crossing, lane changes, merging).

Each participant cycled their own bike, and we provided each with a reflective jacket and helmet. HRs were measured continuously throughout the experiment using a 'Suunto memory belt', a lightweight plastic belt which is worn around the chest and detects HR by means of two thin electrode strips making contact with the chest. The participants also wore a 'Sports Action Helmet Head Camera' which was secured by a strap around their head and pointed directly in front of their face in order to capture their frame of view. The camera can be seen in Fig. 12.2. At scheduled locations along each route, participants were instructed to announce a number between 1 and 10 to represent the degree to which they felt at risk, based on the risk rating scale in Table 12.3a. A risk rating of 1 would denote 'Very Little Risk', and a risk rating of 10 would denote 'Risk of Severe Accident'. The participants, who were all residents of the study area, were informed of the locations of the scheduled rating points before beginning the routes. The audio recording of the head-mounted camera captured the risk ratings announced by the participants, and the video recording allowed the authors to identify the location and context associated with the risk rating. Since the real-time clocks of both the HR monitor and video camera were synchronised, the data from both could be related (Table 12.3b).

Data Analysis and Results

An example of the HR and risk rating data collected is shown in Fig. 12.1. As the figure shows, there were significant variations in both HR and risk rating throughout the test routes. In order to investigate the relationship between HR and risk rating, we divided the HR data into groups based on the associated risk ratings. A box and whisker plot of the HR data grouped by risk rating is displayed in Fig. 12.2. The plot appears to reveal a linear trend between HR and risk rating, but the pattern is not well-defined. We further investigated the relationship between HR and risk rating by performing analysis of variance (ANOVA) of the HR data, considering risk rating as a treatment. In order to account for differences in subjective risk assessment of

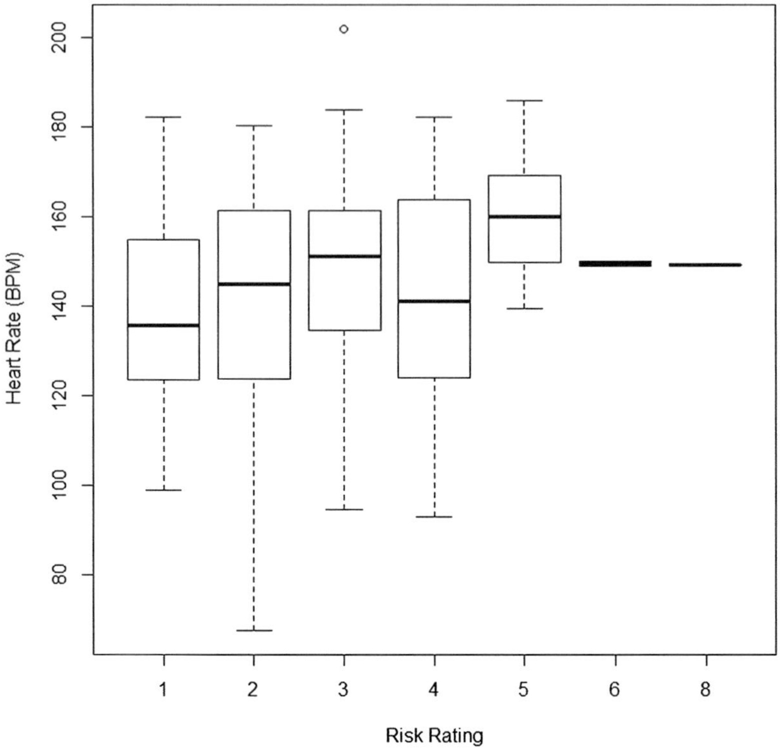

Fig. 12.2 Distribution of heart rates grouped by associated risk rating in the partially controlled experiment

Table 12.3a Risk rating scale for partially controlled and uncontrolled experiments

Perception	Very little risk			Moderate risk					Risk of severe accident	
Rating	1	2	3	4	5	6	7	8	9	10

Table 12.3b Culpability rating scale for diary incidents

Perception	Other completely at fault			Equally responsible					Self completely at fault	
Rating	1	2	3	4	5	6	7	8	9	10

different subjects, we also considered the identity of the individual cyclist as a treatment. Accordingly, two-way analysis of variance was carried out, considering the reported risk ratings and the individual participants as treatments and the HR recorded at each rating point as the repeated measure. No risk ratings of 7, 9 or 10 had been recorded, and the data pairs with risk ratings of 6 and 8 were discarded as their variances were trivial. The results of the two-way ANOVA showed a significant dependency between HR and risk rating ($F(4,141) = 2.7$, $p = 0.03$), indicating that

Table 12.4 ANOVA table for the results of the partially controlled experiment

Source of variation	Degrees of freedom	Sum of squares	Mean square	F-ratio	p-value	Significance
Risk rating	4.00	4114.00	1028.00	2.77	0.03	*
Participant	7.00	40290.00	5756.00	15.50	0.00	***
Residuals	165.00	61253.00	371.00			

Table 12.5 Student-Newman-Keuls test results

Risk rating	Mean HR	Set
1	139.43	B
2	141.41	B
3	148.63	A:B
4	142.22	B
5	160.5	A

the variance in HR between the risk rating groups was significantly greater than the variance in HR within those groups. Therefore, it is possible to reject the null hypothesis that the difference in average HR between the risk rating groups was merely due to chance (i.e. that risk rating and HR are independent of one another). This does not necessarily infer that as risk rating increases, so does HR, but it does show that the HRs for at least two of the risk rating groups appear to come from populations with statistically different means. There was also a significant dependency between HR and the individual participant ($F(7,141) = 15$, $p = 1.54e\text{-}14$) as would be expected. However, there was a non-significant interaction effect of the two treatments, risk rating and individual participant, on HR ($F(24141) = 0.77$, $p = 0.77$). Therefore, the null hypothesis of no interaction—that the differences in HR under the categories of one treatment are not different for any two categories of the other treatment—cannot be rejected (Kirk 1995). What this indicates is that, although HR was dependent on both the individual and their risk rating, the systematic differences in HR associated with different risk ratings were similar among different individuals. As the interaction term was insignificant, we repeated the two-way ANOVA without interaction effects, and the results for the risk rating and participants were the same as before. The full results of this final ANOVA are shown in Table 12.4.

In order to show which levels of risk ratings were statistically different from one another in terms of the heart rate they elicited, a post hoc test was needed. We used Student-Newman-Keuls (SNK) method to perform pairwise comparisons between the means of the risk rating groups. At the 5% significance level, SNK divided the risk rating groups into two sets which were statistically different from one another, one containing the risk rating groups 1, 2, 3 and 4 and the other containing the risk rating groups 3 and 5. The results are shown in Table 12.5. This implies that the mean of risk rating group 5 was significantly higher than the means of risk rating groups 1, 2 and 4 at the 5% significance level. The inclusion of risk rating group 3 in both sets indicates that its mean was not statistically different from the means of either set.

Uncontrolled Study

Experiment Design

For the uncontrolled experiment, surveys were distributed to 39 volunteers, including students and staff of UCC as well as members of the public service. Twenty completed surveys were received from male volunteers between the ages of 22 and 24. In order to complete the survey, participants needed to complete their normal cycling routine over the course of 1 week and to complete a log entry in a cycling diary after each journey. We advised the participants that each diary entry should include any noteworthy incidents such as accidents, near accidents, collisions of any kind or any other incidents which caused a notable reaction or change in risk perception. For each incident reported, we asked the participant to give a risk rating on the 10-point scale in Table 12.3a, ranging from 'Very Little Risk' to 'Risk of Severe Accident', and a culpability rating on the 10-point scale in Table 12.3b, ranging from 'Other Completely at Fault' to 'Self Completely at Fault'.

The participants of the uncontrolled experiment also filled out a road element type survey. This required them to characterise their regular route into a sequence of road elements. We limited the choice of road elements to a list which was adapted from one used in a previous study (Parkin et al. 2007). We advised participants that not every possible road element type was included in the list and so not every element on their own route necessarily needed to be reported. Participants then gave a risk rating for each road element on the 10-point scale in Table 12.3a.

Data Analysis and Results

In total, there were 57 incidents recorded, 38 of which were deemed relevant to the study. We systematically inspected the recorded incidents with the aim of finding common themes. We found that the recorded incidents could be divided into 12 incident types. Table 12.6 shows the frequency and the average risk rating and presumed culpability rating for each incident type. The standard deviation for the average risk ratings was 1.5. The incident types with the highest risk rating were 'Near head-on collision', 'Car door opened in front of cyclist' and 'Car turning right across cyclist at junction'. The standard deviation for the average culpability ratings was 0.7, and, on average, the cyclists admitted the highest culpability for incidents with pedestrians.

Table 12.7 summarises the results of the road element type survey. The road element types with the highest average risk ratings were busy roads without a bicycle lane (5.08) and roundabouts where the cyclist continued straight on (5.04). The road element type with the lowest average risk rating was off-road cycling path (1.06). The standard deviation of the average risk ratings was 1.12.

Table 12.6 Summary of diary incidents

Incident type	Frequency	Average risk rating	Average culpability rating
Car overtaking closely	5	6.8	2.2
Car pulled out in front of cyclist	5	5.0	2.2
Car turning right across cyclist at junction	5	8.0	1
Car parked/stopped in cycle lane	4	4.0	1
Incidents with pedestrians	4	4.3	3
Incidents with cyclists	4	4.8	1.5
Car door opened in front of cyclist	3	8.0	2
Car overtaking at speed	2	5.0	2
Car sideswiped cyclist	2	7.0	1
Poor road surface	2	4.0	1
Car turning left, cut across cyclist	1	6.0	1
Near head-on collision	1	8.0	1

Discussion

The main finding of the partially controlled experiment is that there is a significant dependency between perceptions of risk among cyclists and HR. The frequencies of risk ratings above 5 were too low to make any inferences about the HRs associated with these risk ratings. However, as shown in Fig. 12.2, for risk ratings of 1–5, higher risk ratings were generally associated with higher HRs. The observed link between risk rating and HR is confirmed by the two-way ANOVA with interaction of HR against risk rating and individual participant. The SNK post hoc test produced a more conservative result, suggesting that the only statistically significant difference in means at the 5% significance level was between risk rating group 5 and risk rating groups 1, 2 and 4. This implies that only high-risk perceptions have a significant effect on HR. However, it is possible that the insignificant differences in HR between the lower risk ratings are simply due to a lack of statistical power and that a larger study with more participants and/or risk rating locations would reveal significant differences.

When we examine the results of the travel diaries, we find that, of the 38 relevant incidents recorded, 28 (approximately ¾) involved interaction with a motor vehicle, compared with 4 each for interactions with pedestrians and other cyclists. Almost all of the recorded incidents (and in particular those with the highest average risk ratings) suggest some fault on the part of a motor vehicle driver. Also, the experiment participants' average culpability rating across all incidents was very low at 1.72. This suggests that not only are motor vehicles perceived as being the greatest source of risk but that drivers of motor vehicles are perceived as being almost exclusively to blame. This is in congruence with previous studies on risk perceptions of cyclists (Lawson et al. 2013). This perception may also be realistic: a study by

Table 12.7 Summary of road element type survey results

Road element type	Frequency	Average risk rating	SD of participant averages
Busy road without bicycle lane	76	5.08	2.24
Busy road with bicycle lane	13	3.7	2.25
Busy road with bus lane and bicycle lane	8	2.5	1.64
Roundabout, straight on	9	5.04	2.55
/with cycling facilities	0	–	NA
Intersection, NO traffic signals, straight	29	4.48	1.91
/with cycling facilities	0	–	NA
Intersection, traffic signals, right turn	14	4.3	2.32
/with cycling facilities	6	3.9	1.14
Intersection, NO traffic signals, right turn	18	4.02	2.35
/with cycling facilities	0	–	NA
Roundabout right, turn	1	4	NA
/with cycling facilities	0	–	NA
Roundabout, left turn	6	3.5	2.35
/with cycling facilities	1	2	NA
Intersection, traffic signals, straight	32	3.36	1.78
/with cycling facilities	2	3.5	NA
Intersection, traffic signals, left turn	13	3.33	1.87
/with cycling facilities	2	4.5	0.71
Intersection, NO traffic signals, left turn	19	3.17	1.46
/with cycling facilities	0	–	NA
Residential street with parking	29	2.53	1.73
/with cycling facilities	0	–	NA
Residential street without parking	20	1.89	0.93
/with cycling facilities	0	–	NA
Mini-roundabout	4	1.75	0.5
/with cycling facilities	0	–	NA
Off road cycle path	8	1.06	0.14

Johnson, Charlton, Oxley and Newstead (2010) which considered video footage from the perspective of cyclists found that drivers were at fault in 87% of notable incidents. We also observed that intersections are a high-risk road element for cyclists. Over half of the recorded incidents occurred at an intersection. Incidents of type 'Car turning right across cyclist at junction' had the highest frequency and highest risk rating of the study.

The road element type survey showed that the road element with the highest average risk rating of 5.08 was 'Busy road without bicycle lane' which was also the most frequently reported road element. In contrast, 'Busy road with bicycle lane' and 'Busy road with bus lane and bicycle lane' had respective risk ratings of 3.7 and 2.5. This demonstrates that the presence of cycling facilities such as bus lanes and bicycle lanes on busy roads can significantly reduce the insecurity felt by cyclists.

Previous studies on road infrastructure and cyclist safety suggest that this perception is justified as on-road bike lanes have consistently positive safety effects for cyclists (Reynolds et al. 2009). The further reduction in average risk rating produced by the presence of a bus lane as well as a bicycle lane may be due to the increased separation from car traffic. The lowest average risk rating of 1.08 attributed to 'Off road cycle path' further emphasises that segregation from traffic considerably increases the security felt by cyclists in a mixed mode network. This corroborates the findings of a previous study which indicated that cyclists prefer separated routes (Winters and Teschke 2010).

The road element with the second highest average risk rating of 5.04 was 'Roundabout where the cyclist continues straight on' (without cycling facilities). Although evidence shows that roundabouts reduce motor vehicle collisions by 30–50% (Elvik 2003), this result would suggest that they are detrimental to perceptions of safety among cyclists. This perception is consistent with Moller and Hels (2008) which found that cyclists perceive the situation in which they are circulating in a roundabout and a car is exiting the roundabout particularly dangerous. Moller and Hels (2008) also found that perceptions of risk among cyclists are significantly higher in roundabouts without a cycle facility. There is evidence that these perceptions are justified in a 2009 literature review which found that multi-lane roundabouts can significantly increase risk to cyclists unless a separated cycle track is included in the design (Reynolds et al. 2009).

The following limitations should be considered when interpreting the results of this study. The study group was composed entirely of young males living in Cork city. The homogeneity in the study group allowed us to focus on this particularly prominent user group and prevent differences between subjects from dominating the results. However, for this reason, it is not known if a similar link between risk rating and HR would be observed in a group of different physical or social characteristics. Future studies may remedy this by using a more diversified study group, given that in these studies, we believe we have developed and validated a methodology which would make a larger study worthwhile. Another limitation lies in the possibility that risky activities tend to involve more exertion. If situations which were perceived as being higher in risk tended to occur during moments of greater exertion, the observed differences in HR between different risk rating groups could have been confounded by the effects of exertion. It might be that the use of other sensors, which might quantify external factors such as gradients, wind and speed, would increase the ability of the method to discriminate between exertion levels and risk perceptions. As the results show, even without this additional quantification, HR and perceived risk are clearly associated. Although these limitations would need to be addressed for detailed quantification of the relationship between risk perceptions and HR, this study was successful in demonstrating that the link exists and developing a methodology which may be replicated in larger studies. An important aspect of the developed methodology is in the significant advancement of the capability of the investigation carried out as compared to what was individually possible by the engineers or the psychologists. It is unlikely that without a STEAM approach, such methodologies can be developed this successfully. For development of similar methodology or advancement of the developed method reported in this chapter, clearly defined goals and understanding of skills and disciplinary tools related to such goals are important.

Conclusions

At the level of multidisciplinary collaboration, this research was important to all participants in quite fundamental ways. We learned to frame research problems in such a way that the potential contribution from everyone to answer the question could be identified. We ensured that the research was informed by common goals. And crucially, the project encouraged lasting communication and helped to break down boundaries to communicate effectively. We feel this experience has equipped us with the tools to take part in other more complex projects, involving inter- and transdisciplinary exchanges and moving on from the 'additive' model we used here, in which there was a role for each team member, into the 'synthetic' model (Defila and di Giuliu 2017, 333), which leads to new knowledge being produced as a result of the collaboration.

At the level of the research question, this research has produced evidence that the heart rates of cyclists while cycling in a mixed mode urban network are linked to their subjective risk perceptions. We have demonstrated that situations which are perceived by cyclists to be high in risk are likely to elicit higher heart rate responses than situations which are perceived to be low in risk. The dependency is most significant when comparing the highest risk ratings to all others but may still exist when comparing lower risk ratings to one another. Changes in heart rate in response to perceptions of risk while cycling may also be accompanied by other physiological responses such as release of adrenaline, but further research would be required in order to establish this. We have also examined the risk perceptions of cyclists in relation to both the road elements they encounter on their regular commute and any specific incidents which caused a notable reaction or change in risk perception over a 1-week period. Busy roads and roundabouts without cycling facilities were perceived as most dangerous, while facilities which separated cyclists from traffic greatly reduced risk perceptions. Most recorded incidents involved motor vehicles, and the low culpability ratings of the experiment participants suggest that drivers were perceived as being at fault.

This research raises concern over the viability of using heart rate to measure training intensity and energy expenditure, particularly during activities which are perceived as being significantly high in risk. It can be postulated that the confounding effects of risk perceptions on heart rate-based exercise management tools are likely to be greatest when cyclists are in regular interaction with car traffic. This research also indicates that an important step to be taken in improving the perceived safety experience of cyclists would be the introduction of more dedicated cycling facilities which protect cyclists from motor vehicle traffic, particularly on busy roads and at roundabouts. Finally, the results encourage efforts such as the Road

Safety Authority's 'Cyclists – We All Share The Road' campaign which is aimed primarily at educating drivers on their responsibility towards cyclists as vulnerable road users (RSA 2013).

References

ACSM (2011) Selecting and effectively using a heart rate monitor, from http://www.acsm.org/docs/brochures/selecting-and-effectively-using-a-heart-rate-monitor.pdf

Åstrand PO (1976) Quantification of exercise capability and evaluation of physical capacity in man. Prog Cardiovasc Dis 19(1):51–67

Åstrand PO, Rodahl K, Dahl H, Stromme SB (2003) Textbook of work physiology: physiological bases of exercise. Human Kinetics, Champaign

Bassett DR, Rowlands A, Trost SG (2012) Calibration and validation of wearable monitors. Med Sci Sports Exerc 44:S32–S38

Cacioppo JT, Berntson GG, Larsen JT, Poehlmann KM, Ito TA (2000) The handbook of emotion, 2nd edn. Guilford Press, New York

Cannon WB (1929) Bodily changes in pain, hunger, fear and rage: an account of recent researches into the function of emotional excitement. D. Appleton and Company, New York

Cavill N, Kahlmeier S, Rutter H, Racioppi F, Oja P (2008) Economic analyses of transport infrastructure and policies including health effects related to cycling and walking: a systematic review. Transp Policy 15(5):291–304

Crouter SE, Churilla JR, Bassett DR (2008) Accuracy of the Actiheart for the assessment of energy expenditure in adults. Eur J Clin Nutr 62(6):704–711

Defila R, di Giuliu A (2017) Sidebar: managing consensus in inter and transdisciplinary teams: tasks and expertise. In: Frodeman R, Klein JT, Pacheco RCS (eds) The Oxford handbook of Interdisciplinarity, 2nd edn. Oxford University Press, Oxford

Dill J, Voros K (2007) Factors affecting bicycling demand – initial survey findings from the Portland, Oregon, region. Transp Res Rec 2031:9–17

Doorley R, Pakrashi V, Byrne E, Comerford S, Ghosh B and Groeger JA (2015) Analysis of heart rate variability amongst cyclists under perceived variations of risk exposure. Transport Res F: Traffic Psychol Behav 28:40–54

Elvik R (2003) Effects on road safety of converting intersections to roundabouts – review of evidence from non-US studies. In: Operational effects of geometrics 2003: Highway operations, capacity, and traffic control. Transportation Research Board Natl Research Council, Washington, DC, pp 1–10

Garber CE, Blissmer B, Deschenes MR, Franklin BA, Lamonte MJ, Lee IM et al (2011) Quantity and quality of exercise for developing and maintaining cardiorespiratory, musculoskeletal, and Neuromotor fitness in apparently healthy adults: guidance for prescribing exercise. Med Sci Sports Exerc 43(7):1334–1359

Hughes RG, Harkey DL (1997) Cyclists' perception of risk in a virtual environment: effects of lane conditions, traffic speed, and traffic volume. Paper presented at the Conference on Traffic Congestion and Traffic Safety in the 21st Century

Johnson M, Charlton J, Oxley J, Newstead S (2010) Naturalistic cycling study: identifying risk factors for on-road commuter cyclists. Ann Adv Automot Med/Ann Sci Conf Assoc Adv Automot Med 54:275–283

Kahlmeier S, Racioppi F, Cavill N, Rutter H, Oja P (2010) "Health in All Policies" in practice: guidance and tools to quantifying the health effects of cycling and walking. J Phys Act Health 7:S120–S125

Keegan O, Galbraith J (2005) Attitudes of cyclists and car commuters to cycling in Dublin. Paper presented at the Paper presented at Velocity Conference, Dublin

Kirk RE (1995) Experimental design : procedures for the behavioral sciences. Brooks/Cole Publishing Company, London

Klobucar MS, Fricker JD (2007) Network evaluation tool to improve real and perceived bicycle safety. Transp Res Rec 2031:25–33

Landis B, Vattikuti V, Brannick M (1997) Real-time human perceptions: toward a bicycle level of service. Transp Res Rec 1578:119–126

Landis BW, Vattikuti VR, Ottenberg RM, Petritsch TA, Guttenplan M, Crider LB et al (2003) Intersection level of service for the bicycle through movement. In: Pedestrians and bicycles 2003: safety and human performance. Transportation Research Board Natl Research Council, Washington, pp 101–106

Lawson AR, Pakrashi V, Ghosh B, Szeto WV (2013) Perception of safety of cyclists in Dublin City. Accid Anal Prev 50:499–511

Leden L, Garder P, Pulkkinen U (2000) An expert judgment model applied to estimating the safety effect of a bicycle facility. Accid Anal Prev 32(4):589–599

Lerner JS, Gonzalez RM, Dahl RE, Hariri AR, Taylor SE (2005) Facial expressions of emotion reveal neuroendocrine and cardiovascular stress responses. Biol Psychiatry 58(9):743–750

Levenson RW (1992) Autonomic nervous-system differences among emotions. Psychol Sci 3(1):23–27

Mesken J, Hagenzieker MP, Rothengatter T, de Waard D (2007) Frequency, determinants, and consequences of different drivers' emotions: an on-the-road study using self-reports, (observed) behaviour, and physiology. Transport Res F: Traffic Psychol Behav 10(6):458–475

Moller M, Hels T (2008) Cyclists' perception of risk in roundabouts. Accid Anal Prev 40(3):1055–1062

National Roads Authority (2012a) Estimating AADT on national roads. NRA, Dublin

National Roads Authority (2012b) Expansion factors for short period traffic counts. NRA, Dublin

Noland RB (1995) Perceived risk and modal choice – risk compensation in transportation system. Accid Anal Prev 27(4):503–521

Parkin J, Wardman M, Page M (2007) Models of perceived cycling risk and route acceptability. Accid Anal Prev 39(2):364–371

Polar (2013) Polar sport zones for cycling. Retrieved 15 November 2013 from http://www.polar.com/en/training_with_polar/training_articles/maximize_performance/cycling/polar_sport_zones_for_cycling

Reynolds CCO, Harris MA, Teschke K, Cripton PA, Winters M (2009) The impact of transportation infrastructure on bicycling injuries and crashes: a review of the literature. Environ Health 8:19

RSA (2013) Cyclists- we all share the road. Retrieved 15 November 2013 from http://www.rsa.ie/RSA/Road-Safety/Campaigns/Current-road-safety-campaigns/Cyclists/

Rutter H, Cavill N, Racioppi F, Dinsdale H, Oja P, Kahlmeier S (2013) Economic impact of reduced mortality due to increased cycling. Am J Prev Med 44(1):89–92

WHO (2011) Action plan for implementation of the European Strategy for the Prevention and Control of Noncommunicable Diseases 2012–2016. Retrieved 02 December 2013 from http://www.euro.who.int/__data/assets/pdf_file/0003/147729/wd12E_NCDs_111360_revision.pdf

Winters M, Babul S, Becker HJEHJ, Brubacher JR, Chipman M, Cripton P et al (2012) Safe cycling: how do risk perceptions compare with observed risk? [Comparative study; research support, non-U.S. Gov't]. Can J Public Health 103(9 Suppl 3):eS42–eS47

Winters M, Davidson G, Kao DN, Teschke K (2011) Motivators and deterrents of bicycling: comparing influences on decisions to ride. Transportation 38(1):153–168

Winters M, Teschke K (2010) Route preferences among adults in the near market for bicycling: findings of the cycling in cities study. Am J Health Promot 25(1):40–47

Chapter 13
PELARS: A Case Study of Collaborative Working in Transdisciplinary Teams

Donal Healion, Sam Russell, Nina Valkanova, and Raffaella Rovida

Abstract The intention of this chapter is to provide guidance and advice to prospective partnerships intending to carry out transdisciplinary STEAM related research. Based on the experiences of the authors during the course of a 3-year European Union (EU) funded project, it seeks to outline the benefits and challenges that arise from working in a diverse group of twelve partners drawn from academic, corporate and non-profit sectors, located across nine countries. The chapter presents the aims and outcomes of the Practice-based Experiential Learning Analytics Research and Support (PELARS) project, along with a brief outline of the funding context and attendant structural and organisational frameworks. It conveys a realistic view of both the opportunities and obstacles encountered during a project of this nature and relays useful information about establishing and maintaining effective working relationships between project partners from different disciplines, each with their own area-specific approaches and methodologies. The authors note that the topics discussed relate to their experience only and are not intended to be read as an exhaustive 'how to' checklist when approaching transdisciplinary research in STEAM projects or applications, but rather as a guide to highlight some of the potential issues that can be encountered on the journey and methods to address these.

Our contribution operates at the edge of two main strands in this publication. On one hand it presents a research project that aims to contribute to STEAM education through technological and non-technological outputs. On the other hand, the project consortium itself is a STEAM team, combining different working methodologies at the intersection between creative design practice and design-driven research with scientific, computational and experimental approaches. The work described has been carried out as part of the PELARS project. (PELARS was classified as a small- or medium-scale focused research project (STREP) and was funded under FP7-ICT-2013-11 Objective ICT-2013.8.2, technology-enhanced learning, under the

D. Healion (✉) · S. Russell
National College of Art & Design, Dublin, Ireland
e-mail: healiond@staff.ncad.ie

N. Valkanova · R. Rovida
Copenhagen Institute of Interaction Design, Copenhagen, Denmark

© Springer International Publishing AG, part of Springer Nature 2019

A. de la Garza, C. Travis (eds.), *The STEAM Revolution*,
https://doi.org/10.1007/978-3-319-89818-6_13

work programme objective: ICT FP7-ICT-2013.8.2-b learning analytics and educational data mining. The project was awarded in late 2013, commenced in February 2014 and lasted for 3 years, finishing at the end of January, 2017.). This research and design project seeks to understand how students learn whilst engaged in open-ended collaborative problem solving (CPS) tasks during practice-based learning (PBL) activities. The main aim of the project is to create a learning analytics system (LAS) and incorporate this into a specifically designed learning environment suitable for implementation in three learning contexts, second-level science, technology, engineering and mathematics (STEM) subjects, and third-level interaction design and engineering education.

Keywords STEAM; Collaborative; Transdisciplinary; Learning analytics; Design research

Background

The PELARS project aims to gain a greater understanding of the learning process in the contexts outlined above by developing a characterisation of students' behaviour and progress during practice-based, inquiry-led and experiential learning activities. Through the use of technology embedded in the learning environment, the project created a LAS to capture both ambient and user-generated data in order to establish the patterns of learning created by students working in small groups whilst engaged in open-ended, hands-on activities during STEM and interaction design education. By comparing these patterns, the PELARS consortium seeks to understand how students learn and to create different tools and methods to integrate this information into the classroom to support teachers and students. With such a varied range of user groups and stakeholders, a design-led user-centred approach was implemented to establish user requirements through contextual research employing a variety of design ethnography methodologies.

The PELARS consortium is detailed in Table 13.1 below. The concept for the project was generated by Arduino and Malmö University, and the funding application was driven by the Copenhagen Institute of Interaction Design who also coordinated the project. The consortium was composed of seven academic institutions, three small to medium-sized enterprises and two non-profit organisations. In this respect, the PELARS project bridged not only different disciplines (design, computer science, pedagogy) and partnership typologies but also institutional visions and approaches from a broad explorative and speculative perspective to a more impact-oriented viewpoint focussed on exploitation.

Table 13.1 PELARS project partners and locations

No	Name	Abbrev.	Country
1	Copenhagen Institute of Interaction Design	CIID	Denmark
2	Universität Bremen	UB	Germany
3	Arduino	ARD	Sweden
4	Universitatea din Craiova	UCV	Romania
5	Danmarks Tekniske Universitet	DTU	Denmark
6	Perch Dynamic Solutions Ltd.	PERCH	Ireland
7	National College of Art and Design, Dublin	NCAD	Ireland
8	Scuola Superiore di Studi Universitari e di Perfezionamento Sant'Anna	SSSA	Italy
9	Malmö Högskola (Malmö University)	MAH	Sweden
10	University College London	UCL	UK
11	European Network of Living Labs	ENoLL	Belgium
12	Fundació Privada pel Foment de la Societat del Coneixement (Citilab)	CTL	Spain

Project Outputs

The PELARS project created an educational environment with embedded technology to allow the analysis of learning activities taking place within that environment. It is composed of four main elements, the learning analytics system, the programmable kit, the learning environment and the visualisation.

The LAS consists of a computer vision system, machine learning processor and mobile documentation tools. The LAS captures students hand and facial movement and their interaction with the Arduino programmable kit hardware and software and integrates this data with input from a mobile device-based user documentation system and a user 'sentiment' feedback device (Fig. 13.1). This data is processed, and the results are graphically represented to students and teachers via an online visualisation.

The programmable kit, created by Arduino, is a novel physical computing kit and development environment that allows the exploration of physical computing concepts in a playful way. By plugging modules (such as LED's, switches, potentiometers, motor controllers, etc.), together students can build interactive devices, control their inputs and outputs via a 'drag-and-drop' graphical user interface enabling them to quickly test the efficacy and functionality of their ideas or theories with minimal assembly time (Fig. 13.2).

The learning environment features modular and mobile educational furniture in which the relevant technology is embedded. The furniture is designed to foster communication and interaction (peer to peer and student to teacher) during collaborative hands-on activity and documentation of ideas and allows for ease of movement and postural changes (Fig. 13.3). Trial results have shown that the physical form of furniture elements within the learning environment has a bearing on group formation

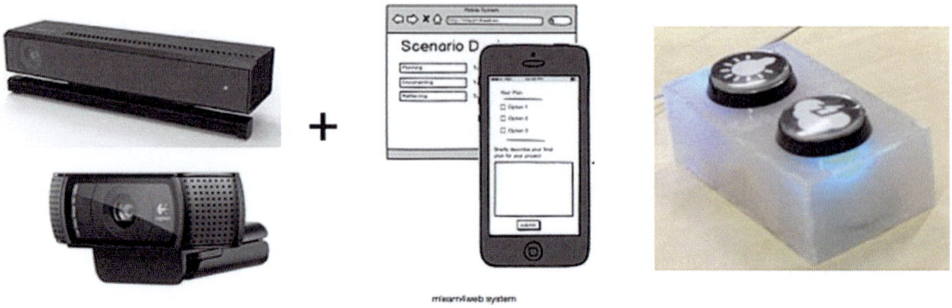

Fig. 13.1 PELARS LAS components featuring computer vision system cameras, a mobile documentation tool and user feedback buttons

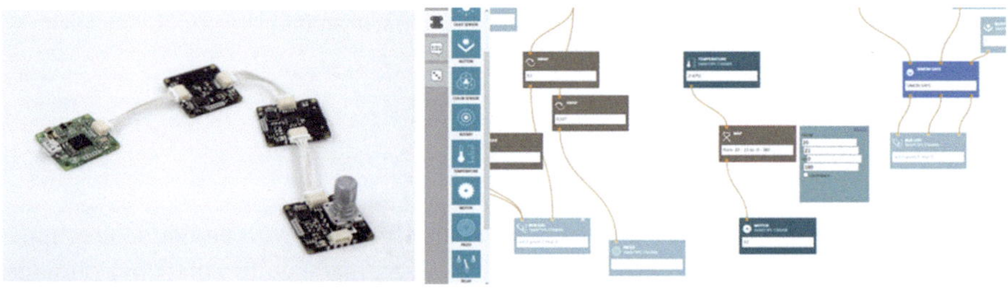

Fig. 13.2 Arduino Talkoo programmable kit and graphical user interface

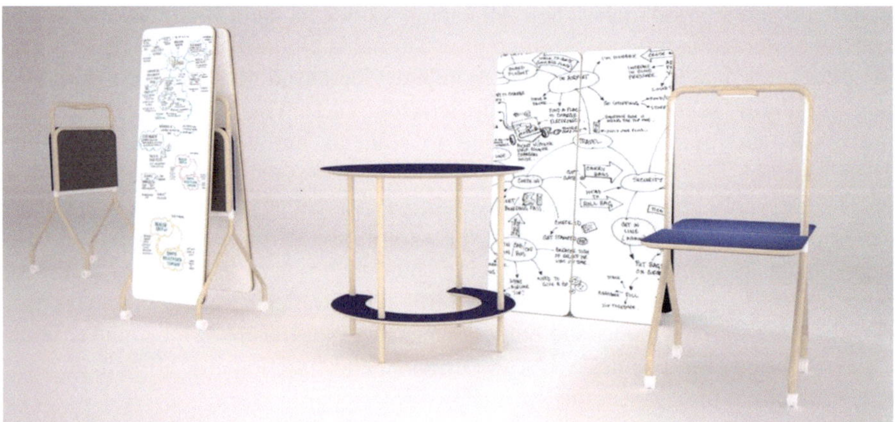

Fig. 13.3 PELARS educational furniture featuring modular, mobile components

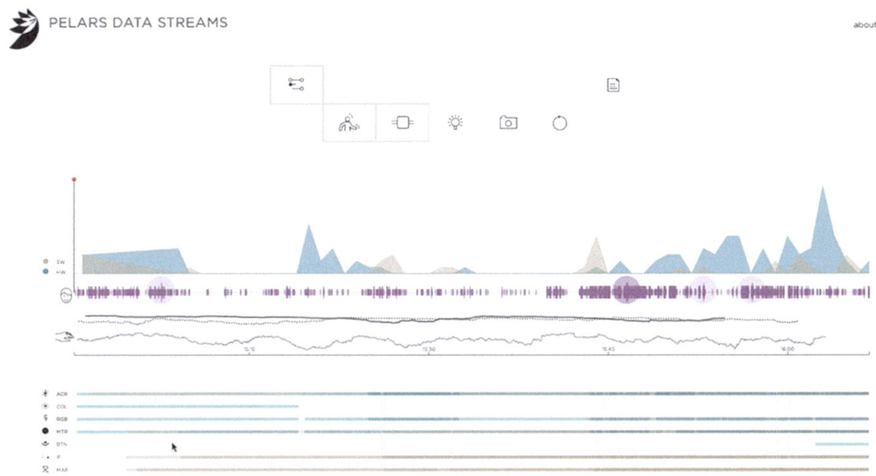

Fig. 13.4 PELARS visualisation of the activity during a physical computing session

and dynamics (Healion et al. 2017). All collaboration is a product of social interaction so by providing standing height, round tables students can work physically closer together and move more easily than if seated at rectangular tables, which facilitates greater interaction and thus collaboration both amongst group members and between groups.

The visualisation is an online interactive graphical representation of the data analysis outputs which provides a deeper understanding of the learning activity process in order to support reflective learning (Fig. 13.4). It enables interactive, layered exploration of the different visualisation components based on students' data recorded and analysed by the LAS during a work session. It provides a timeline-based rich multimedia contextualisation of the visualised components, consisting of image and video thumbnails that are generated by the students as part of their documentation process or by the system as an automated documentation support feature. It also offers an additional summary view, which visualises students' interaction with and connection pattern of programmable kit components, a free-form conversational format to support a language-based storytelling mode, and more advanced analytics such as comparisons of different physical engagement behaviours like hand and face movement and proximity.

Project Outcomes

In order to analyse the implications of the data generated by each of the above components, the Nonverbal Indexes of Students' Physical Interactivity (NISPI) framework was generated. This framework focusses on nonverbal student interactions from the data collected automatically with the LAS and uses students' head direction and hand movement data to interpret key constructs of collaborative practice-based learning including synchrony, individual accountability, equality (symmetry of contribution) and intra-individual variability. The results show that the NISPI coding scheme can be used to observe and interpret these concepts in practice-based learning environments and contributed to the research agenda of multimodal learning analytics with technologies such as PELARS. As such, the consortium added knowledge in the learning sciences field by introducing a new methodology to interpret the key concepts of equality, individual accountability and synchrony as well as bringing in the concept of intra-individual variability as a potential indicator of the quality of collaborative learning in practice-based learning activities (Spikol et al. 2017).

Reflecting the mix of academic and commercial partners, exploitable outcomes of the project consist of both scientific dissemination and commercialisation. To date there have been 24 publications documenting the research carried out in the PELARS project ranging from academic papers and posters to peer-reviewed journal articles and book chapters, as well as policy workshops, exhibitions, pop-up events and conference presentations. Although the project has finished, scientific dissemination is continuing to report on the results of ongoing analysis from data gathered during the latter stages of the project. From a commercialisation viewpoint, project partners Arduino will launch the Talkoo programmable kit, developed during the project on the market along with an associated graphical user interface (GUI). Perch Dynamic Solutions Ltd. are developing a range of commercial educational furniture, based on research findings during the project regarding the efficacy of standing height, round tables as a means to encourage interaction and collaboration during small group work in hands-on type activities.

PELARS Structural Context

Each EU Framework Programme[1] (e.g. FP7 or Horizon 2020) is divided into work programmes organised by sectors and themes which are further divided into funding calls. Typically, prospective interested parties respond to a public funding call by

[1] PELARS was funded under the Seventh Framework Programme (FP7) for research and innovation, a component of an ongoing initiative by the EU to foster collaboration and knowledge transfer between academia and industry in order to increase growth and employment in member states. FP7, now closed and superseded by Horizon 2020, ran from 2007 to 2013, although some of the last projects awarded under FP7 are still running at the time of writing. Whilst FP7 was initially a research-focussed initiative, modifications were made to the programme structure over its 6-year duration to reflect the Innovation Union approach (an initiative within the Europe 2020 strategy to increase economic growth through innovation) which is the basis of the current Horizon 2020 programme (EU Commission, 2010). As PELARS was one of the last batches of FP7 projects awarded, the focus of the programme was already moving towards this research to innovation approach with an emphasis on the commercialisation of any intellectual property generated during the project.

creating a project consortium and developing a funding application. The application contains all the stipulated information regarding the project intent, required resources, partner capabilities, work to be completed and potential impact of the project to allow the application to be evaluated by the EU in a competitive, point-based process. The successful applicants then sign a grant agreement with the EU which contracts the consortium partners to carry out the work detailed in the funding application. The work to be carried out, the project organisation and structure along with all the agreed resourcing allocations are contained in a document called the 'Description of Work' (DoW). The DoW formalises the delivery of the work contained in the funding proposal document and lays out the structure of work over the duration of the project. The overall body of work is divided into thematic- or topic-based work packages (WPs). These are further subdivided into work package tasks which are specific sequential pieces of work that collectively develop a solution to, or answer, the research question posed by the WP outline. WP deliverables document or describe the research work completed in the tasks, typically in report format, but can also be prototypes, demonstrations or other formats (e.g. websites or videos) and are formally assessed by the EU-appointed project reviewers in an annual technical review process.

Whilst each work package is led and managed by one specific partner (work package leader) who has particular expertise in the relevant topic, most of the WP tasks and deliverables are shared with other project partners who contribute to their completion (with the exception of the management WP which is primarily conducted by the project coordinators). As some of the tasks and deliverables are interdependent across WPs, it is in all partners' interest to complete their work to schedule to ensure integration of work across WPs and forward momentum in the project. Each WP also contains a number of milestones which serve as target dates by which to achieve a certain body of work, specific development or understanding and consolidate the work carried out within the WP. The work to be carried out in PELARS was divided into eight work packages listed in Table 13.2.

Table 13.2 Work packages within the PELARS project

WP number	WP title
WP 1	Project Management
WP 2	User Experience Research and Iterative Prototyping in Real Learning Environments
WP 3	Educational Furniture, Environments and Infrastructure
WP 4	Hardware & Lab-ware Kits for STEM Learning & Analytics
WP 5	Learning Analytics Data Acquisition, Reasoning & Presentation
WP 6	Learning Activity and Curriculum Design
WP 7	Trials in Real Learning Environments & Impact Evaluation
WP 8	Dissemination, Community and Communications

Project Organisation

The project coordinators (CIID) were responsible for the overall project management, coordination of WPs, deliverables and meetings, risk management and the establishment of an advisory board and project committees and were the main contact point for financial and progress reporting to the EU project officer. In general, the intention of the coordinators was to keep the project organisation structure as light as possible, in order to maximise efficiency and utilisation of resources towards the realisation of research work. Three committees and an advisory board were created to oversee different aspects of the project. The project steering committee, composed of the work package leaders (WPL), was responsible for the smooth day-to-day running of the project, handling administrative and contractual matters (i.e. contract amendments), and coordination of technical, budgeting and communication between the working groups. The ethics committee was responsible for creating, implementing and documenting templates and procedures to ensure that user research and trials were conducted in an ethical manner and that any user-related data generated was used, handled and stored in an ethical manner in adherence to European, national and institutional guidelines. Compliance with these procedures was checked by the ethics managers (UCL, CIID). The quality committee, led by the quality managers (MAH, CIID), was responsible for ensuring the scope, methodology and standard of research carried out was at the required level. This committee also organised and oversaw an internal review procedure to quality control all the deliverables produced during the project before submission to the EU project officer.

An advisory board (AB) was established to engage a wide set of stakeholders experienced in the fields of education and technology to ensure research relevance, compliance with standards, maintain links to external developments in areas related to the research programme and the diffusion of project results. The advisory board was comprised of the representatives of organisations that are active in educational research and industry and/or have been involved in related EU projects.

Project Challenges

With the organisational and structural elements in place, the focus of the consortium turned towards conducting the research, design and development work required to produce the deliverables outlined in the DoW. However, the formation, organisation and maintenance of a large and diverse STEAM research team can pose several challenges—especially when constraints due to funding structures are ready. These are outlined and discussed in the sections below.

Discovering User Needs: Human-Centred Design Research in EU-Funded Projects

In human-centred design practice, design teams generally undertake contextual research at the beginning of a project. This typically consists of fieldwork with users and in-context research interventions (including methodologies such as cultural probes, design games and co-creation sessions). These methodologies allow the designer to gain insights into user behaviour and help define opportunities for design solutions that may provide the most impact and value to users. This research aims not only to identify unmet user needs but more broadly to unfold new opportunity spaces in which it is possible to frame the challenge of the project from a new perspective. This often results in an open-ended discourse allowing the emergence of further innovative research questions and the reshaping of any preconceived mental or application models before subsequent phases of concept generation and prototyping are carried out.

However, as already stated, EU-funded projects are typically defined by open calls and granted on the basis of a description of their final outcome in all the detail that would bring this outcome to fruition. This sequence makes it difficult to apply human-centred design processes as the structure itself in which EU projects are conceived allows very little exploration of uncertainty through fieldwork or, more importantly, it mostly excludes any deviation from a project plan for which the grant was awarded. The EU granting framework can in this way prevent projects from leading to innovative new applications, interests and research. As it is, due to structural, design and resource constraints of EU research funding instruments, only those user needs which were pre-identified at the time of proposal writing can be addressed within the project. Once a grant is awarded, the application of a human-centred design process in such constraining settings can end in frustration. Consortium partners commit to the project proposal with their expertise, often building on their previous research achievements and outputs, and some may find human-centred design processes distract from individual research agenda goals, making collaboration difficult and even conflictive. Even if fieldwork may reveal interesting, possibly new, places in which to ground the project, the consortium may be prevented from pursuing these directions by the very format of the programme supporting the research.

The PELARS partnership encountered this problem in particular at the beginning of the project. Contextual user research was scheduled to run in parallel with technical project developments (3D object recognition technology and use of radio-frequency identification) that were expected to open up novel ways of understanding and analysing the hands-on STEAM process. However, these technologies turned out to be unsuitable for the application which meant a re-evaluation and redirection of effort. Such preconceptions, as well as the overall structural constraints in this context, impeded the consortium from keeping an open mind to novel, unexpected or latent needs (and frustrations) of stakeholders in the STEAM context that could point to truly valuable, learner- or teacher-centred development strategies. As a result, a strong awareness was raised amongst the consortium that certain flexibility was necessary to allow for shifts in the project development that may come along with discoveries from user research, unexpected technological constraints or research hypotheses that needed to be revised.

Establishing an Agile and Iterative Development Process

As the initial project structure (determined by the DoW) did not allow for agile strategic decisions, the PELARS first year process and plan, as reported to the EU Commission at the Year 1 Technical Review, resembled a 'plan-implement-deploy' waterfall model. This model was rightfully identified by the reviewers as not only slightly outdated but potentially harmful for the project and advised (in the Technical Review Report) the adoption of a more agile and iterative approach with repeated cycles of prototyping, testing and evaluation to run throughout the remainder of the project. The consortium sought to address this challenge by defining an overarching timeline that was independent of individual work packages. This consisted of a series of prototype and user feedback cycles that sought to integrate each of the partners' work outputs in prototype form, position these in the relevant real-world educational contexts, test their functionality and seek user feedback. This approach supported the testing and validation of each partner's research whilst also pushed towards more effective integration of each of the elements of the project. The increased provision of a series of integrated tests also acted as a constructive meeting point for research teams across the project and supported greater collaboration. However, in reality, introducing this approach during the project rather than at the start meant that it was still limited by DoW commitments and varying partner resourcing across the project, which was cumbersome to reallocate once agreed.

Balancing Project Goals with Partner Aspirations (Stakeholder Alignment/Prioritisation of Tasks/Decision-Making)

In order to ensure success at a project level, it was important to endeavour to achieve stakeholder alignment and to facilitate task prioritisation and decision-making towards the common goal of the project in parallel with local aspirations. Whilst minimal formal structures were put in place to address this, the adoption of a more agile project plan and the development of inter partner working relationships supported improved alignment as the project progressed. Regular communication and an evolving understanding of different disciplines, associated methodologies and technologies supported task prioritisation and group decision-making. The rotation of general assembly meetings between the project institutions allowed for a better shared understanding of each of the partner research groups, their resources, working context and goals. Alongside this, the development of a series of prototyping and user feedback cycles that required cooperation between partners and functional integration of physical and digital outputs provided the impetus for each of the stakeholders to align with the overarching timeline, aims and goals of the project—along with the requirement to implement recommendations from the technical review process in subsequent work.

Understanding People in STEAM: Theory Versus Ethnography

To address the methodological or structural discrepancy of integrating design-driven approaches into EU-funded ICT projects, the project consortium discussed different ways of generating insights, establishing requirements and subsequently identifying design opportunities. The project consisted of a transdisciplinary team composed of learning scientists, engineers, computer scientists and interaction and product designers (amongst others). Researchers in the different disciplines sought to generate insights on how a STEAM environment works, or does not work, in quite different ways. On the one hand, learning scientists would typically study and 'project' literature findings on practice-based and project-based learning into an established list of criteria that aim to support the research and design of novel STEAM interventions within the project framework. On the other hand, design researchers would typically want to engage in fieldwork, conduct in-depth interviews and observe the teachers and learners and possibly organise workshops from which the STEAM team (the project consortium itself) can distil valuable insights. Skewing research only into one direction was not useful and felt frustrating to both 'factions': The learning scientists found that relevant knowledge has already been 'discovered' and documented by experts and were not eager to dedicate much time or resources to support field work activities. Design researchers, however, felt that an ethnography-based approach, where connection and trust are established between the researcher/designer and user groups, would deliver subtle and nuanced information that would lead to valuable insights and generate truly useful ideas. In practice, a blend of approaches was achieved, where information and insights gained through contextual research informed and input into the development of theoretical frameworks to analyse and evaluate the proposed design solution.

Observation Is Not Enough: Using Experience Prototyping to Imagine and Discuss

As we create new types of educational tools and environments—especially ones powered by new kinds of technology—and respond to new behaviours, we have to test them in novel ways. The design partners in the project conceived that we were not designing a piece of technology, furniture or a specific tool but rather that we were designing how novel concepts fit into the world (in this case an open-ended, hands-on learning STEAM environment). Accordingly, a key part of the process embraced by the design partners was to try and put themselves in the mind-set of a person that is going to use the resultant design—and to 'use' it in context. In design practice, this approach is called 'experience prototyping', and it is one way to test out the overall experience rather than just the 'thing' being designed. The basic idea is to create a rough prototype (often using simple mock-ups and hacks of off-the-shelf technologies) and then go through the process of using it in its intended environment. This enables the design team to stand in the future, experiencing a new concept without needing to go through the expense of building it.

Throughout the PELARS project, the design partners in the consortium have undertaken several sessions of experience prototyping, probing the experience of different technological and environmental concepts within real-world STEAM contexts. Feedback from stakeholders also enabled new conversations with technological and learning science partners regarding the capabilities and limitations of applying the PELARS system for learning assessment in such STEAM environments, in particular regarding the actual implications for the teachers and their tutoring and assessment practice.

Bridging the Transdisciplinary Gap

Each discipline has a tendency to use domain-specific language, based on a shared understanding and knowledge of certain acronyms, terms and background theories. This day-to-day language is used almost unconsciously between peers and can therefore lead one to the assumption that everyone understands the same ideas and vocabularies within a given topic. The challenge within transdisciplinary research teams is to establish a common understanding of the 'language' of each discipline and the key concepts, theories and approaches that each discipline normally employs in order to carry out their work. The intent is to create a greater understanding of the critical elements of each discipline as they relate to the project at hand, amongst all partners, across each discipline.

This 'bridging' of the linguistic and knowledge gap between each partner discipline is crucial to establish a common baseline understanding amongst all partners of the factors affecting each individual partner's area of work (and how it relates to the project), the position that each partner takes in relation to these factors and how they intend to respond to those factors during the course of the project. Obviously, it is advantageous to establish this understanding early in the project so that there are no misconceptions about what is being discussed, agreed or worked on by individual partners.

Soft Skills

The PELARS project was a human-centred research and design project conducted by humans. Therefore, many of the factors that affected the operation of the project were associated with human relations and dynamics. Although consortium partners may have collaborated on the project proposal, the reality of large multi-partner funded research projects is that many of the partner organisations may not have worked together previously and may not be familiar with other partners, their competencies or their role within the project. Although costly and time-consuming given the typical geographical spread of partner organisations, the importance of face-to-face meetings is not to be underestimated as these create an opportunity to establish working relationships and to organise how the collective work will be carried out where issues can be resolved and decisions made more quickly than over electronic means of communication where a lack of tone, inflection and body language can give rise to misunderstanding. As with any group of randomly gathered people, a transdisciplinary team of researchers will be composed of a mixture of extrovert and introvert personalities. Therefore, the equitable chairing of discussions and meetings is vital to create a balanced and supportive atmosphere where the contributions of those who may not feel confident voicing opinions, particularly outside of their main discipline or in a non-native language, can be sought and encouraged.

As many of the tasks and deliverables across the work packages are interdependent, completion of work on time by all partners is crucial to the overall project running to schedule. Therefore, provision of timely progress updates and responses to requests for information by other consortium partners is an obvious but important way to maintain good working relationships between partners. Within the PELARS project, there were approximately 20 staff changes within partner organisations, with some organisations seeing a complete change of staff by the end of the project. Whilst this is to be expected given the 3-year duration of the project and the involvement of 12 partners, it does reflect the need to ensure good communication between partners to maintain continuity and that measures are in place to allow the smooth handover and transition from one staff member to another.

Language and Cultural Barriers

Whilst language and cultural barriers were not particularly evident within the consortium (English being primarily the de facto language to communicate between partners), there were language barriers experienced during the contextual user research and trials phases. Initial user research was carried out mainly in Spain and Romania and consisted of a number of interviews (conducted by the local project partner) with high school teachers, teaching assistants and students and also observation of technology classes. Whilst the results of these interviews were translated from the original language and communicated within the consortium, the language barrier did prevent a certain amount of in-depth questioning (by non-native partners) during the interview process, and possibly some of the subtleties in responses were lost in translation. The language barrier, in tandem with resource limitations for travel to multiple trial locations, was in part responsible for the fact that the rich set of user groups that were involved in the contextual user research were perhaps not utilised to full potential within the project. In hindsight, follow-up demonstrations and trials of the design iterations with the user groups that had been involved in the research phase would have been an interesting mechanism within the project to trace the response to the development of the system design from initial research and insights through the prototype refinement to the final design.

Recommendations and Lessons Learned

Organisational

In hindsight, the scope of the PELARS project proposal was too wide. Whilst an examination of the three educational contexts had merit, in practice, there were considerable challenges posed by the diversity of the contexts as well as the differences in terms of national education systems, curricula and cultural aspects. An in-depth evaluation of the practical consequences of the project scope would have been beneficial during the proposal drafting process.

To enable effective transdisciplinary work, a clear research, design and development methodology needs to be defined, shared, understood and agreed during the proposal development. Focused workshops involving the whole partnership would be of benefit in order to codesign critical shared elements and jointly validate research findings and project milestones. In the case of PELARS, the design-research methodology was not fully understood or embraced by the whole partnership due to differences in disciplines and approaches and the lack of a clear methodological process plan in the DoW. This led to an incomplete sharing of information and group understanding of knowledge and a partial overlapping of activities and duplication of effort in terms of contextual research, prototyping and validation.

The WP deliverables and associated timelines would have benefitted from a more flexible and responsive structure to better support an iterative, user feedback-informed development process that incorporated all elements of the project. Integration of the work and results of the various WPs was intrinsically hindered by the fragmented WP structure of the project enabling each partner to explore their field of research and pursue their scientific or commercial goals without bringing them back to address the common challenge and impact of the overall project. In effect, from one aspect, the WP structure and specificity in the outcome definition prevented the exploration of valuable solutions not considered in the proposal, and from another aspect, the lack of structure in terms of challenge, goal and impact definition allowed a missed opportunity to bring the partnership onto a higher understanding and sometimes even to a common understanding. Therefore, the inclusion of a specific integration work package including tasks such as conceptual integration and technical integration would have been beneficial to bring together each partner's work on complex and diverse components of the project that ranged across several WPs in a more coherent and coordinated manner.

Involvement of a larger and wider range of user groups at an earlier stage to allow validation through Living Labs could have provided smoother integration through iterative feedback sessions. The engagement of a more diverse range of stakeholders (such as accreditation bodies, European advisory board of policymakers) in the validation process may have provided valuable information to the sustainability, curriculum and policy aspects of the project. Also, the alignment of the general assembly meetings in the latter half of the project with other significant activities (e.g. user trials, exhibitions, Maker Faire, etc.) was a positive development and would have benefitted the project if introduced earlier.

Technical

The complexity and amount of work involved in some of the tasks was underestimated and therefore under-resourced within the project. This limited the relevant partners' capacity in terms of research, design, development and analysis and led to a mismatch between the expected results and resources allocated. Also, the requirement and importance of some specific competences were overlooked at the time of proposal writing. Specifically, a partner specialising in multichannel audio would have complemented the computer vision system work and expanded the range of information available for multimodal analytics. As it was, the abilities (and limitations) of the computer vision tracking technologies determined the relevant input for the learning analytics data sets. The practicalities of the technology meant that object tracking was not possible and the range of data produced became more limited than originally intended. In hindsight, a more coherent approach would have been to identify the pedagogical aims, the information and metrics required to achieve those aims and then develop the technologies and interventions to acquire this information. However, this would be difficult to achieve in the structure of concurrent work packages when development time is limited.

The cost of each LAS and work station and the logistics involved in their production and delivery was prohibitive to the ability to run trials simultaneously and created barriers for the organisation of pop-up events and the engagement of partners in the realisation of user trials and validation. Clear identification and allocation of relevant partner budget for the production of prototypes would be beneficial or alternatively, a central fund, administered by the project coordinator, for the production of prototypes and logistics of running trials. Also, greater collaboration and planning of trial activities amongst the relevant partners to conduct coordinated multimodal data acquisition to produce fully synchronised, precise data sets would have subsequently allowed more in-depth analysis.

Conclusions

Conducting transdisciplinary research can bring many benefits as well as challenges. Creating a project team, building and maintaining working relationships, establishing when, how and by whom work will be carried out all take effort, coordination, communication, persistence and resilience. Fundamental to the process is the recognition that each discipline has its own methodologies, approaches, competencies and language. It could be argued that the success or otherwise of a transdisciplinary team relies on the ability of its members to bridge the knowledge and linguistic gap between disciplines and establish a shared understanding through open communication. The effect of personalities, the development of personal working relationships, sense of humour and the importance of social interaction between partners should not be underestimated in this process.

Allied to the above is the agreement and adherence to the structural framework in which the work is to be carried whilst acknowledging that different approaches may require some flexibility in order to take unforeseen results into account or allow the project to develop along a path that may not have been envisaged at the time of proposal formulation but that emerges through the research and conceptualisation phases of the project.

Works Cited

European Commission (2010) EUROPE 2020: a strategy for smart, sustainable and inclusive growth. COM (2010) 2020 final. Brussels, Belgium

Healion D, Russell S, Cukurova M, Spikol D (2017) Tracing physical movement during practice-based learning through multimodal learning analytics. In: Proceedings of the Seventh International Learning Analytics & Knowledge Conference (LAK '17). ACM, New York, NY, USA, pp 588–589. https://doi.org/10.1145/3027385.3029474

Spikol D, Ruffaldi E, Cukurova M (2017) Using multimodal learning analytics to identify aspects of collaboration in project-based learning. In: Smith BK, Borge M, Mercier E, Lim KY (eds) Making a Difference: Prioritizing Equity and Access in CSCL, 12th International conference on computer supported collaborative learning (CSCL) 2017, vol 1. International Society of the Learning Sciences, Philadelphia

Chapter 14
Geomorphology and Philosophy: A STEAM Survey of the Anthropocene

Valeria Dattilo and Francesco De Pascale

Abstract Many researchers propose that the Anthropocene represents a new division of geological time, positing that our activity by our use of fossil fuels has warmed the planet, raised sea levels, eroded the ozone layer and acidified the oceans. We contend the Anthropocene can only be understood in an interdisciplinary way, integrating ideas from the natural and social sciences with philosophy. That is, by means of STEAM.

Keywords Geoethics · Anthropocene · Kojève · Post-historical epoch · STEAM

The Anthropocene: Contemporary Debate from the Natural Sciences to the Humanities

This chapter considers the new processes of the Anthropocene epoch through the disciplinary lens of geoethics, a sphere in which philosophy, socio-anthropology, geography and the study of geomorphology find confluence. In 2000, Paul Crutzen and Eugene Stoermer of the University of Michigan in Ann Arbor argued that because the industrial agencies of our global population had begun to impact planetary processes themselves, the current geological epoch should be named the Anthropocene (Crutzen and Stoermer 2000, pp. 17–18; Crutzen 2002). The idea inspired many geologists, particularly Zalasiewicz and other members of the Geological Society of London, who were tasked with forming the Anthropocene Working Group to look into the matter (Zalasiewicz et al. 2008). The debate about whether to declare a new geological epoch resurfaced in August 2016 at the International Geological Congress in Cape Town, South Africa. In addition, other scholars are evaluating the issue for the International Commission on Stratigraphy

V. Dattilo
Department of Humanities, University of Calabria, Cosenza, Italy

F. De Pascale (✉)
Italian National Research Council, Research Institute for Geo-Hydrological Protection, Cosenza, Italy

© Springer International Publishing AG, part of Springer Nature 2019

A. de la Garza, C. Travis (eds.), *The STEAM Revolution*,
https://doi.org/10.1007/978-3-319-89818-6_14

(ICS). For an Anthropocene epoch to be added to the official timeline of Earth's history, the backing of the International Commission on Stratigraphy (ICS) will be needed, as well as ratification by the executive committee of the International Union of Geological Sciences (IUGS). The Anthropocene Working Group expects to take 2 or 3 years, at least, to settle on the best golden spike or spikes, the markers for the scientists to point to in millions of years and identify as the geological start of the Anthropocene epoch; the golden spike, in brief, is a physical point in the geological record that shows when one epoch changed to another.

According to Clive Hamilton, the Anthropocene cannot be defined merely by the broadening human impact on nature, which simply extends what humans have been doing for a millennia (Hamilton 2016). The Anthropocene Working Group, which includes Crutzen, initially leaned towards his idea of the Industrial Revolution as its beginning, with other scholars interpreting the Anthropocene as continuing the impact of people on the terrestrial biosphere (Monastersky 2015b, p. 145; Hamilton 2016). As Hamilton indicates, modifying landscape and vegetation may bear the human hallmarks, but these cannot have sufficient impact on the planet to bring about a new geological era. Other scholars argue that the Anthropocene's starting date should depend on when human societies first began to play a decisive role in shaping the Earth's ecosystems (Monastersky 2015b; Hamilton 2016). Also, according to Hamilton, the Anthropocene began when humans changed the functioning of the Earth system (Hamilton 2016). A few scholars include archaeology in the debate, dating the beginning of the Anthropocene to the expansion of agriculture and livestock cultivation more than 5000 years ago or a surge in mining more than 3000 years ago (Ellis 2011). Yet other scholars insist that the Anthropocene is the most recent phase of a process that started 50,000 years ago with human geographic expansion.

In spite of all this scholarly speculation, members of the Anthropocene Working Group have proposed 1945 as the unambiguous point at which people caused a significant shift in the functioning of the Earth system (Zalasiewicz et al. 2014). The first A-bomb test in 1945 contributed to the first stratigraphic presence of radioactive elements in the Earth, and much larger nuclear detonations taking place over the next half century have contributed significantly more (Monastersky 2015a). Other tangible markers of this profound modification may be sought in the alteration of the soil caused by plastic waste, fossil fuel residues, infrastructure and other long-term presence of substances produced by human activity and embedded in the Earth's environment through long-lasting modifications such as the increase in carbon dioxide in the atmosphere (Giorda 2016). Furthermore, in this era known as the 'Great Acceleration', people were increasingly migrating from rural areas to urban centres, feeding the growth of megacities (Monastersky 2015a). These changes put enormous pressure on the environment and the biosphere and underlie the concept of the Anthropocene, literally as 'Man's Epoch'.[1] We are now living in a period in

[1] When we mention 'Man' in this chapter, we mean, as Aristotle said, 'the animal who has language', which is the human capacity to make history, and this is, undoubtedly, a natural datum which distinguishes our species from the others; of course, we don't refer to 'Man' as a male species but to *Homo sapiens*, to human life generally, both female and male.

which our species has become a new geological force capable of impacting the very physical processes of the planet. One of Zalasiewicz's critiques of the Anthropocene as a geological epoch is that compared to other epochs, it is very short, but the Anthropocene Working Group contends that it constitutes such an epoch because many of its changes are irreversible (Carrington 2016).

Therefore, it is not a mere environmental crisis because the term *crisis* refers to something transitional, but we are in a point of no return. Why? Following Bonneuil and Fressoz's reflections, the Anthropocene confronts us with two realities: on the one hand, on the Earth that has existed for 4.5 billion years, life will continue in some form or other and with or without humans. Moreover, even if humans were to drastically reduce their ecological footprint by inventing a sober civilization, the Earth would need hundreds of millions of years to recover the climatic and geo-biological regime of the Holocene. The traces of our urban, industrial, consumer-driven, chemical and nuclear age will remain for millions of years in the geological archives of the planet (Bonneuil and Fressoz 2013).

Consequently, the Anthropocene is not only the subject of contentious debate in the Earth sciences, but it has been appropriated (perhaps uncritically in some instances) by the humanities. However, although much disagreement is engendered by the term Anthropocene, there is a common consensus that it can best be understood through an interdisciplinary approach, by integrating ideas from the natural and social sciences with disciplines in the humanities (Trachtenberg 2016). We therefore advocate a STEAM framework to engage with the challenges of the Anthropocene and seek to mitigate its worst effects. The French sociologist, anthropologist and philosopher Bruno Latour's thesis that Anthropocene is 'the current geological era marked by the impact of human beings on the equilibrium of the planet' was inspired by an article by Richard Monastersky (2015b). The piece, published in the journal *Nature* in 2015, refers to the Anthropocene in terms of climatic, economic, environmental and social changes. His account of disturbing changes occurring from 1950, caused by a huge increase of carbon dioxide production, radioactivity, toxic waste, deforestation and overbuilding, appears to have had a considerable influence on Latour. In a recent essay, Latour positioned the Anthropocene on entirely new foundations and associated the term with social and environmental justice issues in order to design a new geopolitics of recognizing 'planetary boundaries' as political boundaries related to peace processes (Latour 2015). Accordingly, Latour elaborates a notion of politics in a new way, by perceiving the Anthropocene as a 'requiem for the human species' (Hamilton 2010).

Another inseparable aspect of the Anthropocene which pervades the observations of geologists, biologists and researchers in the field of Earth sciences is war. Latour notes that among disasters that have struck the Earth (Gaia), not as a system but as an entity possessing history, war is the 'disaster' par excellence. In this regard, it would then be advisable to try to imagine this epoch, not only as linked to climate change but by practicing a kind of archaeological *epoché* that suspends, at least temporarily, the attribution of the predicates with which we usually define the

Anthropocene. Weapons production and trade that fuel the permanent wars and phenomena like ISIS, the increase of neo-slavery and of irregular migration and the increased gap between wealth and poverty lead to re-situating and redefining the Anthropocene. Therefore, its study encompasses multiple types of phenomena in such a way that only an STEAM interdisciplinary approach can be useful.

The Anthropocene: Geoethical Implications

The principal cause of the Anthropocene is social, rooted in the exceptional capacities of Earth's first ultra-social species: modern humans (Ellis 2016). Therefore, geoethics, a new disciplinary approach which can help us reflect on such issues, now comes into play. Geoethics focuses on how scientists can become more aware of their social responsibilities and how to guide society on matters related to safety in the face of natural hazards, sustainable use of resources and protection of the environment (Peppoloni and Di Capua 2012). Consequently, geoethics is a multidisciplinary approach that encompasses the geosciences, sociology, philosophy, economy and geography. The International Association for Promoting Geoethics (IAPG) has defined the field as 'the research and reflection on those values upon which to base appropriate behaviours and practices where human activities intersect the geosphere' (IAPG 2015). The birth of geoethics occurred when geoscientists became aware of the fact that their activities interfere with and, in some cases, alter in irreversible ways the natural processes of the geosphere. In addition, this field recognizes that scientific choices can have negative consequences on the environment and jeopardize the survival of many species, including humankind (Peppoloni and Di Capua 2015a). By addressing environmental problems at the local and global scale, geoscientists working with social scientists, artists and humanists can help find solutions and ways forward (Peppoloni and Di Capua 2015b). Geoethics encourages a critical analysis of the use and management of natural resources, promoting eco-friendly development; it deals with problems related to the management, communication, education and mitigation of natural risks, by fostering the proper and correct dissemination of the results of scientific studies and information on the risks (Peppoloni et al. 2015). It has been proposed that an ethical pledge similar to the medical 'Hippocratic Oath' be established for geoscientists (Matteucci et al. 2012). In simple terms, geoethics provides guidelines for human behaviour by introducing ethical principles in order to deal with the natural resources of our planet. It guides our use of the Earth's resources for meeting current human needs and their impacts on both environment and society (Limaye 2015). Furthermore, geoethics concerns all the moral questions regarding human impacts on the Earth system (Paal 2015). In addition, Martin Bohle states that our species has acquired the power to engineer planet Earth, be it intentionally, by number, by ignorance or by negligence (Bohle 2015, 2016). Anthropogenic global change is the paradigm of our times and therefore needs to become an explicit part of our value systems (Bohle 2015, 2016). According to Paál (2015), it would be fairly appropriate to state: 'geoethics goes along with the Anthropocene'. More essays are stressing this connection.

Therefore, the Anthropocene requires a sophisticated approach to space, time, knowledge, ethics, politics, social action and, most of all, interactions between human and environmental systems, including the empirical and ontological blurring of these categories (Cook et al. 2015). The processes of Anthropocene can only be understood by integrating the most specific physical and geological field with the human and cultural one.

The Anthropocene: A STEAM Approach

What probably makes Anthropocene a problematic and at the same time a fascinating concept lies in a contradictory statement: Anthropocene could actually reveal the fallacy of anthropocentrism. If the causes of the imminent catastrophe lie in a certain set of human activities, it is obvious that through the *anthropos*, humanity, in general, cannot be considered 'guilty' of poisoning the Earth. As Bruno Latour (2015) pointed out, Anthropocene shakes the very notion of *anthropos*, a universal subject (species, class or multitude) capable of acting as a single people, as a single great individual with its own will. It is as if human society were a collective being that would be the new agent of geohistory, as it happened in the past with the proletariat (Latour 2015).

Latour's hypothesis is that it is absurd to talk about the anthropic origin of global warming, if we consider anthropic as something like human species, without immediately instigating thousands of protests. Even if global warming is anthropogenic in its origin, there is no corresponding 'humanity' that can act under the guise of a single political agent (Chakrabarty 2012). It is not just a question of ethnocentrism, to the extent that it is correlated with anthropocentrism, but rather a question of the exploitation and oppression of the living—that is why the term 'Capitalocene' has also been proposed (Moore 2016; Haraway 2016).

While the term Anthropocene is currently very controversial in geology, since it has to comply with the temporal and stratigraphic limits already mentioned in the previous paragraph, in human and social sciences, however, Anthropocene has worked as a driving force for the development of new research fields which run parallel to the two aspects of the post-human (Marchesini, Haraway, Braidotti), as well as those of the political ecology (starting from Guattari and Gorz) and the Environmental Humanities (Oppermann, Iovino, Holm, Travis, Neimanis, Sörlin) and as a *medium* to strengthen the link between environmental research and socio-political commitment. In addition, the generic reference to *anthropos*, that is to man without distinction, has given rise to a new wave of reflections, theoretical constructions and deconstruction about the relationship between nature and culture, human and non-human, genders, cultures and, more generally, as anticipated, concerning an alleged essence or authenticity of 'man' (Baranzoni et al. 2016). Therefore, the concept of the Anthropocene is an interdisciplinary sharing tool but also a meeting place, milieu or culture broth to create hybridizations between socio-anthropological (Viveiros de Castro, Latour, Avelar), philosophical (Colebrook, Parikka, Stengers,

Stiegler, Szerszynski, Hörl, Haraway, Braidotti) and historical-political perspectives (Chakrabarty, Jason W. Moore).

Anthropocene is presented by some scholars as an event, a point of no return, a 'shock' (Bonneuil and Fressoz 2016). According to Latour, Anthropocene is the most decisive philosophical, religious, anthropological and political concept ever produced as an alternative to ideas of modernity (Latour 2013). Extending the systemic ecology that had included human activities 40 years ago into an analysis of the functioning of ecosystems and biosphere, the idea of Anthropocene represents the dialectical reconciliation between nature and culture, human history and life and Earth history (Bonneuil and Fressoz 2013). It is therefore a change of our relationship with the world.

Anthropocene is political since it implies arbitrating between different human antagonistic forces on the planet, between fingerprints caused by different human groups (classes, nations), from different technical and industrial choices or between different patterns of life or consumption. It is then important to tackle Anthropocene politically to overcome the contradictions and limitations of a model of modernity that has been globalized after two centuries and explore the traces of a rapid and equally divided reduction of ecological footprint of societies (Latour 2015).

The concept of Anthropocene is also fundamental in human geography, whose subject matter is traditionally defined as the set of relationships between man, the environment and society and consequently the study of the changes produced by the processes of interaction between human systems and environmental systems. The introduction of the concept in geographic studies is very recent; Jamie Lorimer's article titled *Multinatural Geographies of the Anthropocene*, published in 2012 on *Progress in Human Geography*, marks an important step in this direction. After noting the end of the concept of nature that has covered the modern era, the article presents the alternative views of environmentalism that focus on the conservation of biodiversity. Through biogeography, these views connect the concept of Anthropocene to the approaches of natural and social sciences (Lorimer 2012). The issue of environmental transformation seems to be able to attract the greatest interest of geographers. Mark Whitehead (2014) places it at the centre of his volume *Environmental Transformations*, subtitled *A Geography of the Anthropocene*, which represents an important attempt to systematize the issue around the themes of resources and life systems. On the one hand, the author presents the state of the situation with respect to the preservation of hydrocarbons, water, air, soil and forests. On the other hand, he observes the role of urbanization and human endeavour to govern environmental change and adapt its approaches to the environment (Whitehead 2014). Three contributions published by Noel Castree (2014a, b, c) on the *Geography Compass* journal, with the aim of exploring the importance of the concept of Anthropocene for the present and the future of geography, should also be explored in order to complete this framework. An additional indication of geography's interest in Anthropocene is the publication in 2015 of a special edition of *Geographical Research* journal. A Symposium of the 2016 Annual Meeting of the Association of American Geographers was devoted to the challenges of

Anthropocene. The 32nd Italian Geographic Congress held in June 2017 also devoted a session to the prospects of Anthropocene and the link with geographic research, geoethics and Environmental Humanities, in which we participated as proponents along with Cristiano Giorda, Paolo Giaccaria and Charles Travis (AGeI 2017). Travis, along with Poul Holm, has published a study on Hannah Arendt's concept of the Polis, detailed in The Human Condition (1958), which addresses the human dimension of climate change (Travis and Holm 2017).

Anthropocene is then linked, on the one hand, to risk perception (De Pascale et al. 2015, 2016, 2017) and environmental limits, with its correlation of emotional and subjective aspects related to the future of humanity; and on the other hand, it is useful to a design component that tries to develop new behavioural responses in the ethical and ecological reorganization of the economy, politics and society (Giorda 2016).

Therefore, there is not only one Anthropocene but many, which overlap and juxtapose in the analysis of researchers who make them their own research object. Or rather, there are only perspectives on the Anthropocene: by maintaining the geographical metaphor that suits the concept of Anthropocene, at least for the time being, only 'cartographies' are possible, factual recognitions that take into account the various positions of the debate while becoming part of it (Baranzoni et al. 2016).

Thinking of Anthropocene requires new Environmental Humanities and a new STEAM approach. It brings the social sciences, the humanities and the natural sciences together in different ways to address the current ecological crises from closely related ethical, cultural, philosophical, political, social and biological perspectives (Oppermann and Iovino 2017). Indeed, this human species that has plunged the planet into the uncertain becoming of Anthropocene is not only a biological entity, but it is also made up of social and ideological systems, of institutions and imagines, of geographies and power relations that hold an irregular distribution of Gaia's benefits and damages, of legitimacy of talking about and for the planet and of possibilities to weigh on technical and economic choices (Bonneuil and Fressoz 2013).

In the next section we analyse the Anthropocene as a symptom of the era we live in, an analysis that contains tropes attributed to the philosophical sphere.

The Anthropocene As a Symptom of the 'Absence of the Future'

The Anthropocene testifies not only to the influence of 'human activity' (Crutzen 2005, p. 54) on the global environment but also provides a prognosis of the fate and destiny of nature itself. Indeed, the irreversible decline, anticipated within such a framing of the Anthropocene, corresponds directly to a 'crisis affecting Man's own being as a political animal' (Crutzen 2005, p. 135). As implied in its subtitle *Man has changed the climate. The Earth enters a new era*, Crutzen's book entitled *Welcome to the Anthropocene!* (2005) concerns research on climate change but elides the problems associated with 'anthropogenic phenomena of this new geological era' (Crutzen 2005, p. 25). This section of the chapter focuses on a philosophical archaeology that asks: 'What is the Anthropocene? What is it about? Does it define and question Man as a political animal? Which anthropological level, in any decisive way, does it include so that every individual can get involved in it? What makes our species historic?'.

The Future As the Original and Founding Time of History

Starting from the last question, it is worth clarifying the meaning of the term *history* as defined by German philosopher Martin Heidegger. In *Sein und Zeit* (1927), his questions concern the concept of history and, specifically, the conditions that make it possible. He states that 'history has its essential weight neither in what is past nor in the today and its connection with what is past, but in the authentic occurrence of existence that arises from the future of *Da-sein*' (Heidegger 1976 ed., p. 462). It is a particular philosophical position, characterized by the primacy of the *future*, that consists in postulating the future as the original and founding time of history. In this sense, the link between historicity and the future is understood not as chronological future, but as an existential future, that is, a future that should be thought of and understood from the perspective of death in relation with the finite. What is at stake here is a finite temporality. This aspect is clearly stated in the second section of *Sein und Zeit*, precisely in the fifth chapter entitled *Temporality and Historicity*:

> History, as a *Da-sein* way of being, has its roots so essentially in the future that death, as the possibility of *Da-sein* we characterized, throws anticipatory existence back upon its *factical thrownness* [...] Authentic being-toward-death, that is, the finitude of temporality, is the concealed ground of the historicity of Da-sein. (Heidegger 1976 ed., p. 462)

Heidegger identifies in the future, in this temporality destined to the end, in what he calls 'being-toward-death', that 'original temporality' that makes our species historic (Heidegger 1976 ed., p. 484). Our species is not only intended to disappear, but it always has a connection with the future as mortality, which means it is constantly connected with the possibility of death.[2] Reflecting on the theme of death, Heidegger suggests that history is a concept that can be summed by four stages:

1. History as deriving from the past
2. History as 'a set of events and effects' taking place in the past, the present and the future
3. History as a body that changes over time and that, distinguishing itself from nature, which moves over time, embraces the events and the fate of men, of human communities and of their culture
4. Finally, history understood as the transmission, be it historiographically recognized or perceived as apparent, although the origin remains obscure

---

[2]According to Heidegger, death is only a simple refection of authentic dying.

In this manner, History appears as the 'occurrence of the existing *Da-sein* over time', with 'Man being the subject of events', of history. This last thesis poses the following problem: 'To what extent and on the basis of what ontological conditions does historicity belong to the subjectivity of the historical subject as its essential constitution?' (Heidegger 1976 ed., p. 458). The problematic of the 'time of species' as elaborated by Heidegger takes on an aphoristic value in *Sein und Zeit*: 'The analysis of the historicity of *Da-sein* tries to show that this being is not "temporal" because it "stands in history", but that, inversely, it exists and can exist historically only because it is temporal at its core' (Heidegger 1976 ed., p. 452). So, if there is no death, there would be no history, since if something existed eternally, the past, present and future would coincide in static nature.

The Post-historical Epoch and the New Epoch of the Anthropocene

From another perspective, Alexander Kojève asserts bluntly that the 'end of history' announced by Hegel has indeed occurred, defining our epoch as post-historical. Attributing our primacy no longer to the future but to the 'eternal present' typical of an environment, Kojève states that we focus on a kind of vanishing and immemorial temporality, in other words, on a type of society that doesn't need to remember its past or project itself towards the future. Rather it is a new type of animality. Post-historical animals, as defined by Kojève, although they keep dying, no longer have any constant or permanent link with the eventuality or the possibility of an end. Rather, they are immersed in a kind of eternal present, typical of non-human animals, which Augustine of Hippo attributed to God:

> Your years are one day, and your day is not any or every day, but Today (*non cotidie sed hodie*) because your Today does not yield to a tomorrow nor did it follow on a yesterday. Your Today is eternity: Therefore, you generated the Co-eternal to whom you said: "This day, I have begotten you. You created all times and you exist before all times. Nor was there any time when time did not exist". (Augustine of Hippo 2006 ed., p. 557)

This rhetorical game played by Augustine of Hippo (2006, p. 557) concerns the concept of eternity in relation to time, in which he discusses the impossibility of thinking of eternity except as a 'motionless and eternal present' (Augustine of Hippo 2006 ed., p. 397). In Augustine's game, time never ends but exists in a non-temporal eternal present from which the three dimensions of time will arise. To parse Augustine through Kojève's argument is to say that our years are similar to God's years, existing all in a single day, a today without yesterday or tomorrow. Without seeking to address here Kojève's philosophy in toto, it will be sufficient and necessary to stress an important point, specifically within Kojève's book *Introduction to the Reading of Hegel* (1947) in order to explain the existing relationship between the Anthropocene and human nature (Kojève 1996 ed.). Commenting on Hegel's *The phenomenology of spirit* (1807), Kojève stated:

> I realized that the Hegelian-Marxist end of history, far from having been ascertained, was already a present fact. Observing what was taking place around me and reflecting on what had taken place in the world since the Battle of Jena, I realized that Hegel was right to see in this battle the end of History properly so-called. With that battle the vanguard of humanity virtually attained the limit and the aim, that is, the end of Man's historical evolution. (Kojève 1996 ed., p. 541; Hyppolite et al. 1980, p. 273)

In other words, what is realized and manifested in the 'end of history' is the depletion of human potential. Or better, the 'not-yet' becomes and takes the form of a 'here and now' of something contingent.[3] The hiatus, the distances separating the infinite from the finite, shortens to the point where they coincide, a gap which is the central point and a decisive one of the history. Man remains without history, projecting himself to a post-human condition, to a stage of new animality. It is in view of the post-human, of a new animality, that one should reread Crutzen's book, *Welcome to the Anthropocene!*. In other words, what takes place and emerges in this epoch is the depletion of humanity potential, the *anthropos* concept.

Now, we can ask ourselves this question: 'What kind of Man lives in the Anthropocene epoch?' This is an important question that Crutzen does not seem to address. Indeed, the world in which we live is divided and dispersed and deprived of past coordinates. This idea of time, as neither short nor long but as an eternal moment, is also present in Zarathustra:

> The walk backward takes an eternity. And the walk forward takes another eternity. These are two opposite directions; they collide against each other: and it is here, at this gate, that they meet. The gate's name is written up there: 'Moment' [...] All truth is crooked; time itself is a circle [...] Look—I continued—this moment! A long eternal road that turns runs from this gate: behind us, there is an eternity. Must not all things which can run have run already on this road? Must not everything which can happen, have happened already, been done with, and flowed away? (Nietzsche 1965 ed., pp. 163–164)

This is the illusion that characterizes our epoch, in particular, our sense of the present in the West, and the feeling that the future is closed, that nothing new happens anymore, and every event that happens is the replication or a copy of something that has previously occurred. For this reason, we feel shaken by the world instead of acting upon it. Hence, there is a link between the problem of an epoch characterized by Man's impact on the environment (Anthropocene) and the change of Man's very nature. According to the analysis of Hegel, Heidegger and Kojève, what will disappear is in fact not only history but Man's *ethos*. This Greek term refers to shared practices and customs, a set of habits that foresees a certain degree of variability. It also infers a certain relationship and dialectic between subject and object (norm and application of the norm in Wittgenstein's terms) and between nature and culture. We can state that 'the only habit to survive is that of no longer having solid habits' and witness, thus, the sunset, the decline of human experience, of the *Erfahrung*, the German term for experience in the sense of tradition, of what is transmittable, characterized by social and historical ties. But there is more, of course:

---

[3]We can also use the following synonyms, *the endless* and *the invariant*, using Chomsky's terms, the *eternal, in* Augustine's terms, and the *language faculty, in Saussurian jargon.*

> The definitive annihilation of Man properly 'so-called' also means the definitive disappearance of human Discourse (*logos*) in the strict sense. Animals of the species Homo sapiens would react by conditioned reflexes to acoustic signals or mimics, and thus their so-called discourses would be like what is supposed to be the 'language of bees'. (Kojève 1996 ed., p. 542)

Nothing is therefore so different from a 'language's instinct', an expression used by Pinker to refer to a type of an innate, natural language, one that doesn't need artifices, that is, a social dimension, a community, not mixed with history. Hence, it is a way of communicating or perceiving, which is not so different from the chirping of birds or from the stimulus-response type, type A produces B, but B reproduces A, that is, a way of behaving which results in a set of automatic and predefined actions, telling us word for word how to act, thus inducing some adhesion of the human animal towards the situations one gradually faces, towards the environment and towards the vital context in which one is requested to act or, better, to react. The language re-environmentalizes becomes again a communication code that is reduced to a simple decoding of the transmitted signals. It is clear then that the idea of an 'end of history' no longer defines only a form of relationship between people based on a certain structure of society but a whole range of existing relations between Man and language, Man and institutions, Man and environment and Man and Man, relationships and relations that are no longer historical or human according to Kojève.

Pulling the Strings

It is now time to ask: 'How do preconceived ideas or opinions and the conditions of possibility change in the Anthropocene epoch? How do subject and object, Man and world, operate now?' We believe that the post-historical animal has a symbiotic relationship with nature and the environment. During the Anthropocene epoch, we witness the manifestations of what are actually the conditions of possibility, the transcendental conditions of historical praxis, in which power takes the form and appearance of the act. These conditions come to light with the new animality.

The theme of new animality is *de vogue*, with recent scholars, such as in the formulation provided in Felice Cimatti's book *Philosophy of the animality* (2013). In this regard, Cimatti identifies in the absolute imminence what Deleuze refers as the real condition of 'becoming animal' of the end of transcendence, of a new human figure with the characteristics of the angel or the infant evoked by Rilke in *Duineser Elegien* (1923). It is at this point that we wonder how is it possible to imagine Man's experience in the Anthropocene, which, according to Cimatti, would be characterized by a state of immanence:

> The world of immanence is a totally new world, a world that is fully here, that aspires to nothing other than to be here, with no "after" or "before," right here. A here that takes a different value as it doesn't project itself either in times that no longer exist or in those that do not yet exist. (Cimatti 2013)

A big aporia that remains, in our view, opens inside the anthropogenic vision of Crutzen.

Conclusion

Based on the foregoing, we can conclude that *history* is considered as 'completed' when 'metahistory', that is, what is subtracted from time, assumes concrete, empirical appearances, reaching a conciliation, an identity of metahistory and history. Furthermore, at the end of this path, we can reassert the relevance of the definition already formulated: reflection on the 'end of history' is not an old refrain, but something current. The idea of 'ending history' was in fact reiterated and deepened in the twentieth century by many scholars, specially by Alexandre Kojève, whom we considered one of the most theoretical radicals of the 'end of history'. He brings, in this way, the Crutzenian anthropogenic vision to extreme consequences and sees, beyond the reflection and formulation of the 'end of history', the relationship between the latter and human nature, between Man and the world in which he is invited to live in.

Contrary to the authors studied, we are firmly convinced that there is no discourse or eternal theory about death, as Heidegger claims in the paragraph that we have analysed, nor on Man as Kojève does, for example, nor on the meaning of history or destiny of the planet as, for example, Crutzen pretends to do, taken up, subsequently, by Zalasiewicz and other scholars such as Steffen et al. (2011) and Bonneuil and Fressoz (2013, 2016).

Claims that are possible only if these categories are considered as the *Da-sein* for Heidegger, or concepts such as history, theories, and categories that are nothing more than *language*, as eternal structures, a priori fixed forever, rather than considering them as *linguistic horizons* neither stable nor eternal, within which the man thrown in, immersed, relates. These are rather temporalized, historically oriented a priori, that is, noneternal a priori. Only in this way can the claim to build an eternal discourse on the meaning of history and its end disappear.

Referring to the discussion on the points of analogy between the authors who have defined *the end of history* and those who support the Anthropocene epoch, we can say that the type of world view of the latter, according to which the planet in which we live has reached a point of saturation, a dead point, or a stalemate situation, so everything has already been done and seen, is not so different from the Kojevian view of history, according to which nothing new would happen in history.

In our view, a different concept of history, world and human nature should be put forward instead; it is then necessary to rethink the relationship between history and metahistory, between time and eternity, between subject and object and between man and the world as two heterogeneous terms in perfect tension between themselves, which coexist with one another without ever reaching a conciliation, an identity.

If we now resume the questions posed at the beginning of the second paragraph—what is Anthropocene? what is it about? does it define and question Man himself as a political animal?—we realize, only now, that they find an answer. Or rather, is it possible to establish an analogy between the 'end of history' and the 'new geological era' analysed by Kojève and Crutzen, respectively? We can say that in neither case is there a need to resort to a metahistorical plan to continue living or rather surviving. With the new geologic era theorized by Crutzen and, in general, by the supporters of the Anthropocene, as well as by the supporters of the end of history, a risk of losing man and, with him, the world in which he lives is highlighted.

This leads some scholars, for example, Bruno Latour, to call upon politics, or a technical system that protects us from the risk of not being present in any possible civil history. In fact, Latour proposes to ecologize instead of modernizing and putting nature into politics through a set of procedures (sometimes scientific and sometimes political) to evaluate the place—irreparably uncertain and controversial—of a multitude of beings in our common world; none of which can serve the simplest means to others (Latour 2012).

Crutzen and Stoermer, on the other hand, propose a worldwide strategy that leads to sustainability of ecosystems and that knows how to use collective intelligence—the noosphere (Vernadsky 1924; Crutzen and Stoermer 2000; Hamilton and Grinevald 2015)—that stems from new ideas and sensitivity towards the environment, landscape and cultural heritage, spreading them globally into a renewed integral ecology that turns into protocols, urban devices and new life cycles: a challenge defined by both scholars as stimulating but difficult and discouraging for the scientific community and the research world.

In addition, promoters of geoethics recognize the contingency of human evolution on the planet (Pievani 2009, 2012), identify *Homo sapiens* as a geological force acting on the geological and biological environments and assign to humans an ethical responsibility that arises from the consciousness of being a modifier of Earth systems (Bobrowsky et al. 2017).

Therefore, if we had to define our epoch, and ourselves within it, we would say that our epoch is erroneously called post-historical or post-modern and we ourselves are mistakenly defined post-historical animals. Our discourse seems to have come to a decisive point: the problem is no longer simply the meaning of the new geological age, of its sense or end, but the most radical one of its own potential that comes from the threshold, from the gap between these two terms, those of end-not end, subject-object, man-world, placing us before it. Probably we don't have to choose between two lines of thought, for example, among those who are in favour of a new geological era and those who are not or among those who support the end of history and those who do not. Rather we need to meditate on the circularity that, indefinitely, uses these two terms interchangeably, and in the repetition of this circle, in its historical possibility, leaving some elliptical displacement to occur: neither a straight line nor a perfect circle.

By echoing Bonneuil and Fressoz's words (2013), living finally means, in the context of the Anthropocene, dwelling in a nonlinear world and unpredictable of Earth's responses, or rather of Earth history, of our perturbations.

We actually think we live, for the first time, in an epoch which allows a complete overlap among geological and geographical categories (e.g. that of the 'Earth system') and certain philosophical concepts (e.g. 'history'), which correspond, respectively, to those of 'nature' and 'culture'. It is, therefore, a plot that would show the man's relationship with the world, becoming an immediate content of social life in a very specific historical period, namely, that of Anthropocene. From these assumptions we can understand the importance of speaking about this geological era in different degrees and from different points of view, which in turn represent several STEAM cartographies of Anthropocene.

Indeed, if the debate about the Anthropocene temporarily stopped in geomorphology, the concept is present in the contemporary collective imagination, and the impact of human factors on environmental changes is quite intense in the social perception (De Pascale et al. 2015, 2016).

Consequently, documenting, understanding and responding to the present and future challenges posed by the recent changes in the relationship between human beings and their environment thus become an imperative for social sciences and humanities.

Author Contributions Dr. Francesco De Pascale is the author of the first section of the chapter, which includes the following paragraphs: *The Anthropocene: Contemporary Debate from the Natural Sciences to the Humanities*, *The Anthropocene: Geoethical Implications* and *The Anthropocene: A STEAM Approach*.

Dr. Valeria Dattilo is the author of the second section which includes the following paragraphs: *Anthropocene as a Symptom of the 'Absence of the Future'*, *The Future as the Original and Founding Time of History*, *The Post-historical Epoch and the New Epoch of the Anthropocene* and *Pulling the Strings*. The conclusions are attributable to both authors.

References

AGeI (2017) Programme and abstracts 32nd Italian geographic congress: "the contribution of geography among revolutions and reforms". Association of Italian Geographers (AGeI), Bologna. ISBN: 978-88-942641-0-4

Arendt A (1958, 1998) The human condition. 2nd edn. University of Chicago Press, Chicago

Augustine of Hippo, *Confessiones* (it. transl. 2006 with Latin text *Le Confessioni*, Rizzoli, Milano, 29th edn). Book 11

Baranzoni S, Lucci A, Vignola P (2016) L'Antropocene. Fine, medium o sintomo dell'uomo? Lo Sguardo, Rivista di Filosofia 22(3):5–9

Bobrowsky P, Cronin VS, Di Capua G, Kieffer SW, Peppoloni S (2017) The emerging field of geoethics. In: Gundersen LC (ed) Scientific integrity and ethics with applications to the geosciences, Special Publication American Geophysical Union, John Wiley and Sons, Inc.

Bohle M (2015) Geoethics: from "blind spot" to common good. In: International Association for Promoting Geoethics blog. iapgeoethicsblogspotit. Available online June 13, 2015

Bohle M (2016) Handling of human-geosphere intersections. Geosciences 6(3):1–11

Bonneuil C, Fressoz J (2013) L'événement anthropocène. La Terre, l'histoire et nous. Éditions du Seuil, Paris

Bonneuil C, Fressoz J (2016) The shock of the Anthropocene: the earth, history and us. Verso, London

Carrington D (2016) The Anthropocene epoch: scientists declare dawn of human-influenced age. The Guardian. Available from: https://www.theguardian.com/environment/2016/aug/29/

declare-anthropocene-epoch-experts-urge-geological-congress-human-impact-earth. [29 August 2016]

Castree N (2014a) The Anthropocene and geography I: the back story. Geograph Compass 8(7):436–449

Castree N (2014b) The Anthropocene and geography II: current contributions. Geograph Compass 8(7):450–463

Castree N (2014c) The Anthropocene and geography III: future directions. Geograph Compass 8(7):464–476

Chakrabarty D (2012) Postcolonial studies and the challenge of climate change. New Lit Hist 43(1):1–18

Cimatti F (2013) Filosofia dell'animalità. Laterza, Roma-Bari

Cook BR, Richards LA, Rutherfurd I (2015) Geographies of the Anthropocene. Geogr Res 53(3):231–243

Crutzen PJ (2002) Geology of mankind. Nature 415:23

Crutzen PJ (2005) Welcome to the anthropocene! Andrea Parlangeli Ed, Milano

Crutzen PJ, Stoermer EF (2000) The Anthropocene. IGBP Newsl 41:17–18

De Pascale F, Bernardo M, Muto F, D'Amico S, Zumbo R, Galea P, Agius M (2015) Neogeography and seismic risk perception. A comparison between two case-studies: Calabria (southern Italy), Malta. Eur J Geogr 6(1):64–83

De Pascale F, Bernardo M, Muto F, Ruffolo A, Dattilo V (2016) Geoethics, neogeography and risk perception: myth, natural and human factors in archaic and postmodern society. In: D'Amico S (ed) Earthquakes and their impact on society. Springer Natural Hazards, International Publishing, Cham, pp 665–692

De Pascale F, Bernardo M, Muto F, Di Matteo D, Dattilo V (2017) Resilience and seismic risk perception at school: a geoethical experiment in Aiello Calabro, southern Italy. Nat Hazards J Int Soc Prev Mitigation Nat Hazards 86(Suppl. 2):569–586. https://doi.org/10.1007/s11069-016-2696-z

Ellis E (2011) Anthropogenic transformation of the terrestrial biosphere. Phil Trans R Soc A 369:1010–1035. https://doi.org/10.1098/rsta.2010.0331

Ellis E (2016) Evolving toward a better Anthropocene. *Future Earth Blog. Research for Global Sustainability*. Available from: http://www.futureearth.org/blog/2016-mar-29/evolving-toward-better-anthropocene. [16 March 2016]

Giorda C (2016) Lo studio dell'Antropocene: una svolta anche per la geografia? Ambiente Società Territorio Geografia nelle scuole 61(3):3–8

Hamilton C (2010) Requiem for a species: why we resist the truth about climate change. Earthscan, London

Hamilton C (2016) Define the anthropocene in terms of the whole Earth. Nature 536:251. Available from: http://www.nature.com/news/define-the-anthropocene-in-terms-of-the-whole-earth-1.20427. [17 August 2016]. https://doi.org/10.1038/536251a

Hamilton C, Grinevald J (2015) Was the Anthropocene anticipated? The Anthropocene Review, pp 1–14

Haraway DJ (2016) Staying with the trouble: anthropocene, capitalocene, chthulucene. In: Moore JW (ed) Anthropocene or capitalocene? nature, history, and the crisis of capitalism. PM Press, Oakland

Heidegger M (1927) *Sein und Zeit*, Halle-Saale (it. transl. 1976, *Essere e tempo*, Longanesi, Milano)

Hyppolite J et al (1980) In: Salvadori R (ed) Interpretazioni hegeliane. La Nuova Italia, Firenze

International Association for Promoting Geoethics (IAPG) (2015) International association for promoting geoethics. Available from: http://www.geoethics.org

Kojève A (1947) Introduction à la lecture de Hegel. Gallimard, Paris (it. transl. 1996, Introduzione alla lettura di Hegel, Lezioni sulla fenomenologia dello spirito, Adelphi, Milano)

Latour B (2012) Enquêtes sur les modes d'existence: une anthropologie des modernes. La Découverte, Paris

Latour B (2013) Facing Gaia. Six lectures on the political theology of nature. Gifford Lectures, Edinburgh

Latour B (2015) Face à Gaïa. Huit conférences sur le nouveau régime climatique, Les Empêcheurs de penser en rond. La Découverte, Paris

Limaye SD (2015) A geoethical approach to industrial development of georesources and groundwater use: the Indian experience. In: Peppoloni S, Di Capua G (eds) Geoethics: the role and responsibility of geoscientists. Geological Society, London, Lyell Collection (Special Publications), vol 419, pp 13–18

Lorimer J (2012) Multinatural geographies of the Antropocene. Prog Hum Geogr 36(5):593–612

Matteucci R, Gosso G, Peppoloni S, Piacente S, Wasowski J (2012) A Hippocratic Oath for geologists? Ann Geophys 55(3):365–369

Monastersky R (2015a) First atomic blast proposed as start of Anthropocene. Nature. Available from: http://www.nature.com/news/first-atomic-blast-proposed-as-start-of-anthropocene-1.16739. [16 January 2015). https://doi.org/10.1038/nature.2015.16739

Monastersky R (2015b) Anthropocene: the human age. Nature 519(7542):144–147

Moore JW (2016) Introduction. Anthropocene or capitalocene? nature, history, and the crisis of capitalism. In: Moore JV (ed) Anthropocene or cpitalocene? nature, history, and the crisis of capitalism. PM Press, Oakland, pp 1–11

Nietzsche F (1883–85) Also sprach Zarathustra Ein Buch für Alle und Keinen (it. transl. 1965, Così parlò Zarathustra, Mursia, Milano, XIV ed.)

Oppermann S, Iovino S (2017) The environmental humanities and the challenges of the anthropocene. In: Oppermann S, Iovino S (eds) Environmental humanities: voices from the anthropocene. Rowman & Littlefield International – Intersections, London/New York

Paál G (2015) Geoethics, the Anthropocene and the pope. In: International association for promoting geoethics blog. Available from: iapgeoethics.blogspot.it. [11 July 2015]

Peppoloni S, Di Capua G (2012) Geoethics and geological culture: awareness, responsibility and challenges. Ann Geophys 55:3. https://doi.org/10.4401/ag-6099

Peppoloni S, Di Capua G (2015a) Introduction. In: Peppoloni S, Di Capua G (eds) Geoethics: the role and responsibility of geoscientists. Geological Society, London, Lyell Collection (Special Publications), 419, pp 1–4

Peppoloni S, Di Capua G (2015b) The meaning of geoethics. In: Wyss M, Peppoloni S (eds) Geoethics: ethical challenges and case studies in earth sciences. Elsevier, Amsterdam, pp 3–14. https://doi.org/10.1016/B978-0-12-799935-7.000

Peppoloni, S, Bobrowsky, P, & Di Capua, G 2015, 'Geoethics: a challenge for research integrity in geosciences', in N Steneck, M Anderson, S Kleinert & T. Mayer (eds), Integrity in the global research arena, World Scientific Publishing Co, Singapore, pp. 287–294

Pievani T (2009) The world after Charles R. Darwin: continuity, Unity in diversity, contingency. Rend Fis Acc Lincei, Springer 20:355–361

Pievani T (2012) Geoethics and philosophy of earth sciences: the role of geophysical factors in human evolution. In: Peppoloni S, Di Caupua G (eds) Geoethics and geological culture. Reflections from the Geoitalia conference 2011, annals of geophysics (special issue), 55, 3, pp 349–353

Rilke RM (1923) Duineser Elegien, Insel Verlag, Frankfurt (it. transl. 1978, *Elegie Duinesi*, Giulio Einaudi Editore, Torino)

Steffen W, Grinevald J, Crutzen PJ, McNeill JR (2011) The Anthropocene: conceptual and historical perspectives. Phil Trans R Soc A 369(1938):842–867

Trachtenberg Z (2016) Bringing conversation to the Anthropocene. FutureEarth blog, Research for global sustainability. Available from: http://www.futureearth.org/blog/2016-aug-24/bringing-conversation-anthropocene. [24 August 2016]

Travis C, Holm P (2017) Lessons for the Anthropocene from the recent past: tobacco use, HIV/AIDS, and social transformation. Global and Planetary Change. Available online 20 April 2017. https://doi.org/10.1016/j.gloplacha.2017.04.006

Vernadsky WI (1924) La Géochimie. Librairie Félix Alcan, Paris
Whitehead M (2014) Environmental transformations. A geography of the anthropocene. Routledge, London/New York
Zalasiewicz J, Williams M, Smith A, Barry TL, Bown PR, Rawson P, Brenchley P, Cantrill D, Coe AE, Gale A, Gibbard PL, Gregory FJ, Hounslow M, Kerr A, Pearson P, Knox R, Powell P, Waters C, Marshall J, Oates M, Rawson P, Stone P (2008) Are we now living in the Anthropocene. GSA Today 18(2):4–8
Zalasiewicz J et al (2014) When did the Anthropocene begin? A mid-twentieth century boundary level is stratigraphically optimal. Quaternary Int. Available online 12 January 2015

Part V
Policy

Chapter 15
STEM to STEAM: Policy and Practice

Marie Clarke

Abstract Higher education institutions have been criticised for not contributing as much as they should to innovation in the wider economy, particularly in their regions (EC, Communication from the Commission to the European Parliament, the Council, the European Economic and Social Committee and the Committee of the Regions on a renewed EU agenda for higher education. Brussels, 30.5.2017 COM(2017) 247 final, pp 1–12, 2017). While STEAM at societal level offers an avenue for the sciences, the arts and the humanities to pool knowledge and resources to address global challenges (de la Garza A, Travis C, The STEAM revolution bridging the divide: transdisciplinary approaches to arts, humanities, science, technology, engineering and mathematics studies, Springer, Cham, 2017), the policy context in which universities operate plays an important role. This chapter argues that in order for STEAM to advance beyond individual initiatives or transdisciplinary/interdisciplinarity debates it requires support in the policy context. Using Kingdon's (Agendas, alternatives and public policies, 2nd edn. Longman, New York, 1995) Multiple Streams Framework (MSF) the policy emphasis on STEM in the Irish context will be explored and the resultant implications for STEAM approaches will be considered. The chapter concludes that current policy approaches in the Irish context makes it challenging for advocates of STEAM to develop the approach as a viable policy alternative.

Keywords STEM · Arts and Humanities · STEAM · Higher Education · Policy

M. Clarke (✉)
University College Dublin, Dublin, Ireland
e-mail: marie.clarke@ucd.ie

© Springer International Publishing AG, part of Springer Nature 2019
A. de la Garza, C. Travis (eds.), *The STEAM Revolution*,
https://doi.org/10.1007/978-3-319-89818-6_15

Introduction

Policymakers are keen to strengthen their country's economic competitiveness and look to innovative education systems to produce the skills and training needed for success in the global economy. Within the higher education context, in addition to developing disciplinary competences, skills such as creativity, critical thinking and the ability to communicate and work in complex problem-solving teams are viewed as vital to economic competitiveness (OECD 2011). Higher education institutions have been criticised for not contributing as much as they should to innovation in the wider economy, particularly in their regions (EC 2017). Policy strategies now combine prioritisation and competition (Dill and van Vught 2009). Innovation policies focus on using market forces emphasising competition for research resources and encouraging universities to pursue multiple sources of funding for research activity (Clancy 2015). While STEAM at societal level offers an avenue for the sciences, the arts and the humanities to pool knowledge and resources to address global challenges (de la Garza and Travis 2017), the policy context in which universities operate plays an important role. In order for STEAM to advance beyond individual initiatives or transdisciplinary/interdisciplinarity debates, it has to become part of the 'policy window' where policy issues move onto the government agenda and result in decision and action (Kingdon 1995).

This chapter focusses on Ireland, which provides an interesting context as the global financial crisis in 2007 resulted in 9 years of austerity measures that impacted on all aspects of the economy. This in turn resulted in policy initiatives that focussed solely on economic recovery and future competitiveness. Using Kingdon's (1995) Multiple Streams Framework (MSF), the policy emphasis on STEM will be explored, and the resultant implications for STEAM approaches will be considered. The chapter concludes that current policy approaches in the Irish context make it challenging for advocates of STEAM to develop the approach as a viable policy alternative.

Policy Analysis: The Multiple Streams Framework

Policy analysis is an area that has generated a range of methodological perspectives. Traditionally the field was dominated by positivist epistemology which viewed policymaking as a linear process from the identification of a problem to choosing a policy solution (Chow 2014). This linear model arising from this approach highlighted a number of challenges that policy analysis scholars needed to address such as the context, the timing and the role played by political ideologies (Chow 2014). The Multiple Streams Framework model developed by Kingdon (1995) tries to reconcile these issues and interprets the policy process as dynamic and irrational due to ambiguity that is present in the real policymaking environment. The framework identifies three different streams: the problem stream, the policy stream and the politics stream that operate in parallel within the policymaking environment (Kingdon 1995). The problem stream identifies the issues that stakeholders and government are conscious of. The policy stream is where solutions are worked out, developed, rejected or accepted. The politics stream refers to public opinion and stakeholder demands that need to be reconciled. Kingdon (1995) argues that when these streams are joined, a policy window develops when the issues become part of the policy agenda and will be enacted in subsequent policymaking. He also identified a group who he called 'policy entrepreneurs' who promote their views and ideas in both the problem and the policy stream and whose actions are key to securing their agendas.

The underlying assumption of the Multiple Streams Framework is that policymaking is dynamic, irrational and unpredictable (Nutley et al. 2007) and the surrounding environment is always ambiguous and complex (Pollitt 2008). Within the framework, ambiguity may be defined as the state of having many ways of thinking about the same problem (Zahariadis 2003). While the framework has limitations such as the lack of attention to the role of the media in policy development (Chow 2014), nevertheless it is a useful lens to explore the issue of STEAM, given the changing perceptions of economic competitiveness and the influence of policy direction in research funding.

STEM to STEAM: The Policy Context

STEM is a central preoccupation of policymakers across the world. The STEM economic policy agenda is largely driven by the need to enhance the general quality of the supply of human capital as well as enlarge the high skill group capable in research, commercialisable innovation and effective response to technological change (Marginson et al. 2013). STEM-strong countries such as China, Taiwan, Korea, Finland and Switzerland have developed strategic national STEM policy frameworks. This is reflected in centrally driven and funded programmes, curricular reform and new teaching standards; world-class university programmes, the recruitment of foreign science talent and new doctoral cohorts; and decentralised programme initiatives and partnerships and engagement that link STEM activities in schools, vocational and higher education with industry, business and the professions (Marginson et al. 2013). In the United States, STEM education was promoted by successive US Presidents as playing a key role in US competitiveness and future economic prosperity. A similar situation developed in the European context where STEM was promoted at each level of the education system and particularly in the higher education context through the determination of research priorities and funding.

While STEM is highly promoted in the EU, the umbrella term STEM is not viewed as a useful category for understanding the supply and demand dynamics in science, technology, engineering and mathematics as it tends to imply a high level of substitution between different education fields and occupations, which is not necessarily possible in practice (EC 2015a, b). Furthermore, there is a lack of agreed statistical definitions within countries and across the EU of what constitute STEM study fields, STEM occupations and STEM sectors. These data gaps mean there is often a lack of adequate data to inform policymaking reliably (EC 2015a, b). At a broad policy level, countries have a broad set of science and technology areas that recur regularly – biotechnology and life sciences, ICT and nanotechnology. Clancy (2015) suggests that governments are not prioritising according to national need but are imitating policy directions in other countries. This has been accompanied by the development of new funding instruments where governments provide less direct funding to universities for research and provide larger proportions of funding to research councils or other funding bodies. Universities are required to compete for these funding streams (Clancy 2015).

The focus of policymakers on innovation and economic development has facilitated a science model of research where impact involves spin out companies, patents and economic return, while 'innovation' is inextricably linked to the notion of technology transfer (West 2013). This is a limited view of innovation and does not account for the mechanisms of public engagement (exhibitions, media appearances, social media, public lectures) and working with cultural institutions and NGOs in addition to collaboration with business, industry and policymakers. From the early 1980s to 2014, European research funding was allocated by a series of multiannual research frameworks, running from I to VII. Collaborative research funding was allocated within disciplinary domain-specific programmes, including health, ICT, nanotechnology, energy, environment and transport. With the introduction of the 4th Framework (1994–1998), social sciences and humanities were granted one of these programmes (DG Research 2009). An analysis of Research and Development Expenditure by Clancy (2015) based on OECD data from 2005 indicates that the level of Research and Development expenditure across 18 EU countries was 70% on science disciplines, whereas the social sciences and arts and humanities attracted a quarter of Research and Development funding. In most of these countries, social sciences received 1.5 times that allocated to the humanities.

Horizon 2020 moved from a disciplinary approach to a focus on grand challenges with an emphasis on European competitiveness (Benneworth et al. 2016). The abolition of the Social Sciences and Humanities Research Directorate of the EC's Research and Innovation Directorate-General, and a greater focus on global challenges and innovation, concentrated attention further on the scientific, medical and (at the margins) social scientific dimensions of European research (West 2013). The heads of European Research Councils (EUROHORCS) and the European Science Foundation (ESF) were abolished in 2011 and replaced by Science Europe. This body has six scientific committees including a separate group for humanities. While humanities continue to have a call on European Research Council funding, there has been a persistent concern that these disciplines are being excluded from global challenge agendas as set by Horizon 2020 in Europe (van den Doel 2013). This perpetuated a view that new scientific discovery, underpinned by the economic, legal and behavioural frameworks, further marginalises the humanities (West 2013).

The shift in emphasis to innovation and creativity on the part of policymakers has led to a renewed focus on the arts and humanities.

According to Maeda (2013) STEAM is about innovation where critical thinkers are developed through education to lead in an innovation economy. Galligan (2014) contends that one of the primary reasons why the arts (and design) have moved onto the national and local policy radar is due to a perceived need to link the arts with the sciences as key education and workforce policy concerns (Galligan 2014). This linkage is referred to as 'STEM to STEAM' which seeks to educate a workforce with creativity, skills and flexibility by encouraging arts integration alignment across domains of college and career readiness, arts education and the academic core (Galligan 2014). Within the research context, STEAM is viewed as promoting innovation within and between the sciences, arts and humanities (de la Garza and Travis 2017). The move to a STEAM approach emerged as a response to a commonly held view that graduates did not possess an innovative spirit which is perceived as a problem in economic development (Land 2013). Supporters of STEAM argue that by fostering creativity, critical thinking, team working skills, and opportunities for personal growth, STEAM education and skill sets enhance students' opportunities for gainful employment and to contribute to societal needs. STEAM is viewed as a cross-curricular collaboration (Land 2013), as integrating the arts into the STEM curriculum providing pathways for personal meaning-making and self-motivation.

STEAM is regarded by critics as a distraction from the strategy of supporting STEM where basic skills are missing and the STEM area requires ongoing resources (May 2015). Other authors such as Ball (2015) argue that creativity is present in science and does not need arts to support its processes. Gulbrandsen and Aastad (2015) argue that a better understanding of innovation and societal progress is required, a debate which will enrich STEM as well as arts and humanities research. While differences remain about the concept of STEAM and what is should achieve, it is clear that it is not fully accepted as part of the policy agenda.

More recently in the United States, STEAM has emerged in the policy domain. In 2013 the US Congress set up a bipartisan STEAM Caucus which held public briefings and worked with STEAM experts. The STEAM movement hoped to capitalise and build on STEM's momentum. In Europe, STEAM has featured more prominently in recent years. A recent EU Report *Science Education for Responsible Citizenship* (EC 2015a, b) acknowledged that greater attention should be given to the value of all disciplines and the ways that interdisciplinarity (STEAM rather than STEM) could contribute to the understanding and knowledge of scientific principles and solve societal challenges (EU 2015a, b). In 2017, the European Commission published a communication entitled *A renewed EU agenda for higher education* which pointed to a need to address the area of high-level skills. It was suggested that in many EU member states, there is an unmet demand for graduates in science, technology, engineering (arts) and math (STE(A)M) fields. High-level digital competences, numeracy, autonomy, critical thinking and a capacity for problem-solving were viewed as increasingly important attributes. The European Commission has committed to launch an up-scaled EU STE(A)M coalition bringing together different education sectors, business and public sector employers to promote the uptake of relevant STE(A)M subjects and modernise STE(A)M and other curricula, including more multidisciplinary programmes and cooperation between relevant faculties and HEIs (EC 2017). It will be up to individual member states of the EU to develop national policies that will create the necessary conditions to realise those objectives.

The Irish Context

The Irish public higher education system consists of seven universities and Dublin Institute of Technology (which have self-awarding powers) and 13 other Institutes of Technology (which have delegated authority to make their own awards from QQI). There are other public and private higher education institutions in specialised areas. Education policy in Ireland is the prerogative of government, and it is vested in the Minister for Education and Skills under the Ministers and Secretaries Act 1924 and the Education Act 1998. The Department of Education and Skills is one of the higher-spending departments in the Irish civil service and its significant budget means that it is of considerable importance not only educationally but also socially. Policy decisions in Ireland are influenced by both the global and more importantly the European context. This is evident in the language employed by various government departments in policy documents, which reflects that of international organisations such as the World Trade Organization, the OECD and the EU (Seery 2008). The pace of economic and social change in Ireland was extremely rapid between 1995 and 2007, and the higher education sector was viewed as central to developing a knowledge economy. The financial crisis of 2007 led to a sudden and sharp decline in funds available to the Irish exchequer. The financial crisis resulted in GDP dropping by 7% from 2008 to 2009. Arising from emergency legislation and national agreements, the total estimated cost reduction in public sector expenditure from 2008 to 2013 was €7.8 billion (DPER 2014), of which a reduction in public sector pay accounted for €3 billion.

The global financial crisis of 2007 and ensuing collapse of the Irish economy led to changed circumstances where policy focus was directed towards the main problem of rebuilding the country's economic infrastructure, focussing specifically on employability. Since 2008 the higher education sector witnessed a series of cuts to funding. In the period 2008–2015, state grants to the sector declined by 38% (Clarke et al. 2015). In the higher education context, the austerity measures were accompanied by a series of policy initiatives designed to make universities more responsive to government policy. Universities were now expected to deliver on national objectives as represented in government initiatives such as *Action Plan for Education 2017*, the *National Skills Strategy 2025*, *Innovation 2020*, and *Irish Educated, Globally Connected*. The future funding of Irish higher education is currently under review. The Department of Education and Skills has indicated that the funding approach will be underpinned by a series of core principles, including the need for it to be metric and outcome based and reflective of national policy priorities (DES 2017a; b). Successive governments recommended a number of reforms for the sector through the publication of policy papers and *The National Strategy for Higher Education to 2030: Report of the Strategy Group* which promoted a view that Ireland needed to be repositioned as an attractive knowledge-intensive economy underpinned by a research-rich but restructured higher education system (Hazelkorn 2014).

STEM to STEAM As a Policy Focus

From the 1960s science was a policy concern in Ireland. A number of reports *Science and Irish Economic Development* (1966) and the OECD (1974) *Review of National Science Policy: Ireland* pointed to the lack of a centralised science policy where individual government departments pursued different initiatives (Clancy 2015). The policy analysis pointed to the need for investment, which did not follow. The National Board for Science and Technology (NBST) was established in 1978 with a remit to advise the government on policy and planning in the area of science and technology and to coordinate public and private investment (Clancy 2015). Policy development was curtailed by the lack of public resources and by the changing nature of organisational structures to support science policies and practices (Clancy 2015). During this period very little research activity took place in the university sector, but by the late 1980s, this had changed, and Irish researchers were successful in securing EU Framework funding. By the 1990s it was clear that a more focussed policy approach to science research and to the role of universities in this area was required (Clancy 2015). In 1995 the report of the *Science, Technology and Innovation Advisory Council* (STIAC) was published which focussed on the development of a national system of innovation to include the enterprise sector, the state sector and the third-level sector. The White Paper on *Science, Technology and Innovation* (1996) was published, setting out government policy in this area. This marked for the first time a unified policy approach to industrial development and science and technology (Fitzgibbon 2011).

In 2000 two research councils, the Irish Research Council for Science, Engineering and Technology (IRCSET) and the Irish Research Council for Humanities and Social Sciences (IRCHSS), were stablished. In 2001 Science Foundation Ireland (SFI) was created. Clancy (2015) has calculated that over the life time of the Programme for Research in Third Level Institutions (PRTLI), €1.213 billion was allocated to research. Over the first four cycles of programme, the sciences secured 82% of total funding: 48% in the bioscience/biomedical area, 27% in physical sciences and technology and 9% in environment and natural resources. Programmes in the social sciences and humanities secured 12% of funding; 6.5% was devoted to two library projects (Clancy 2015).

With the onset of the economic crisis in 2007, the research funding landscape began to change as the government pursued a more economically targeted approach.

The production of new knowledge became core to economic growth (Harpur 2010), in turn raising the profile of STEM research (Hazelkorn 2014). In May 2010, responsibility for PRTLI transferred from the Department of Education and Skills to the (then) Department of Enterprise, Trade and Innovation, and SFI became the principal funding agency. In 2013, the remit of SFI was extended to allow it to fund applied research in line with the government's policy of research prioritisation, which was adopted in 2012. The Research Prioritisation Exercise (RPE) marked the end of what had been a strategy to build a broad base of expertise in favour of a more targeted approach. Research relevance, defined principally in terms of job creation, became the paramount criterion, with an emphasis on science and technology. Fourteen areas, aligned with industrial sectors, were identified (Hazelkorn and Gibson 2016).

While STEM has been emphasised in the policy context, Ireland does not have a specific STEM education policy. *Innovation 2020* Ireland's strategy for research and development emphasised STEM at all levels of the education system as a way of enhancing Foreign Direct Investment (FDI) and creating an enhanced eco system for indigenous companies to develop (STEM Education Review Group 2016). *Innovation 2020* does not identify the absence of a national policy or strategy in this context and does not call for its establishment (STEM Education Review Group 2016). Student underperformance in international assessments such as Trends in International Mathematics and Science Study (TIMSS) and the OECD's Programme for International Student Assessment (PISA) has resulted in a number of government interventions to enhance STEM education. In second-level education, new programmes in mathematics (Project Mathematics) were introduced on a phased basis between 2008 and 2013 as part of Junior and Senior cycles. Concerns about the low uptake of higher-level mathematics at Leaving Certificate level led to the introduction of bonus CAO points. New curricula for Junior Certificate Science and Leaving Certificate Biology, Chemistry and Physics are awaiting implementation, and a new specification in Agricultural Science is in development. The new STEM curricula emphasise the development of skills as well as discipline knowledge and understanding. More recently concerns have been expressed about the underpreparedness of students entering third-level courses with a high level of mathematics content (STEM Education Review Group 2016). There are also concerns about the level of digital skills. National studies highlight that, where ICT is used, it is mainly for low-level activities such as word processing, internet searches and playing computer games. Limited use is made of ICT in the development of higher-order thinking skills, creative or collaborative skills, independent working skills or communication skills (Eivers et al. 2010). A different policy approach was pursued in the arts and humanities areas.

Arts in education in the Republic of Ireland has suffered from fragmentation and compartmentalisation as well as under-resourcing (Coolahan 2016). The Arts Council, the national agency for promoting and developing the arts in Ireland, was established in 1951 and operates under the *Arts Act* 2003. The need for joined-up, integrated collaboration across government departments, education agencies and arts organisations was first addressed in the Arts Council of Ireland Report *The*

Place of the Arts in Irish Education (Benson 1979). The appointment of the first minister for the Arts, Culture and the Gaeltacht (Ireland's Irish-speaking regions) in 1993 demonstrated a government commitment to this area; however, policy development remained disjointed. In 2009/2010 arts was part of the Department of Arts, Sports and Tourism; in March 2010, it came under the remit of the Department of Tourism, Culture and Sports, and since June 2011 responsibility for the arts rests with the Department of Arts, Heritage and the Gaeltacht.

The publication of the *Employment and Economic Significance of the Cultural Industries in Ireland* (Coopers and Lybrand 1994) indicated that the arts had become part of public policy dialogue with reference to economic growth and employment (Hazelkorn 2014).

The publication of *Assessment of the Economic Impact of the Arts in Ireland* (Indecon 2011) pointed to the arts as important to the economic policies that the government had identified as crucial to economic recovery. The arts were viewed as central to foreign direct investment, the creation of an imaginative labour force, and in establishing an innovative environment in which the creative and cultural industries could thrive. The second and third Global Irish Economic Fora (held in 2011 and 2013) highlighted the role of arts and culture as a key vehicle for economic growth and recovery and a means of restoring Ireland's international reputation (DAC and G 2015). The fourth pillar of the *Creative Ireland Programme* launched in 2016 focusses on establishing Ireland as a centre of excellence in media production. The initial focus in 2017 will be on Ireland's potential to be a global leader in film production TV drama, documentary, children's storytelling and screen animation. The programme also provides a commitment to develop an integrated plan to enable every child in Ireland to access tuition in music, drama, art and coding by 2022. As part of this process, more attention is now being paid to the role played by the arts in the education system at all levels.

The *Arts in Education Charter* (2013) co-signed by then Ministers of the Department of Arts, Heritage and the Gaeltacht (DAHG) and the Department of Education and Skills (DES) acknowledged the need for an education system that promoted creativity. Various curricular initiatives at pre-primary level (Aistear early years curriculum, 2009), the Primary School Curriculum (1999) and the Framework for the Junior Cycle (2012), all reinforce the emphasis on creativity and innovation. A number of other initiatives including Encountering the Arts Ireland (ETAI) provide children and young people with quality arts and education encounters; the Arts in Education Portal and the National Arts and Education day all serve to raise the awareness of the arts. A recent study *Arts and Cultural Participation among Children and Young People: Insights from the Growing Up in Ireland Study* (Smyth 2016) has highlighted significant social differentiation in children's cultural participation where those from higher-income families are more likely to read for pleasure and attend after-school music or drama lessons or clubs. The study also found girls' participation in these activities to be higher than boys regardless of age and social background. Children from immigrant backgrounds, especially at primary level and even earlier, were less likely to take part in cultural activities (Smyth 2016). In the higher education sector, there are a number of partnerships between the Arts Council and Higher Education Institutions (HEIs) which have led to the funding of artist residencies.

The HEA report *Playing To Our Strengths* (2010) focussed specifically on the role that AHSS could contribute to the wider economy. It recommended that course development should facilitate undergraduates and postgraduates in developing generic, technical and business skills. Optional internships in the workplace were recommended for all AHSS students where they would experience entrepreneurship and innovation training. Another recommendation suggested that all SET programmes should have AHSS modules specifically designed to provide graduates with relevant skills associated with these disciplines. Some progress has been made across the higher education institutions to address these recommendations, but little policy attention has been given to these areas since the publication of the report.

In 2012, the IRCHSS and IRCSET merged to form the Irish Research Council (IRC). This left no dedicated agency for arts, humanities or social sciences research. While the funding allocated to the arts, humanities and social sciences areas was considerably less than what was given to the sciences, nevertheless a number of capital projects, such as the Humanities Institute of Ireland in UCD (PRTLI 3), the Trinity Long Room Hub (PRTLI 4), Maynooth University's Iontas Building for HSS research (PRTLI 4) and the Arts, Humanities and Social Sciences Research Building NUIG (PRTLI 5), and large-scale inter-institutional collaborative projects, for example, Humanities Serving Irish Society (PRTLI 4), Graduate School of Creative Arts and Media (GradCAM) (PRTLI 4) and the Digital Arts and Humanities Structured PhD programme (PRTLI 5), received funding (Hazelkorn and Gibson 2016).

The Research Prioritisation Exercise has made it difficult for the arts, humanities and social sciences to secure research funding due to the period of fiscal constraint. The reduction in the block grant to universities which provided support for research in all disciplines and the distribution of research funding to priority areas has led to a situation where disciplines that cannot easily demonstrate their direct relevance to economic growth run the risk of becoming marginalised (Clancy 2015).

STEAM as a term has not featured prominently in Irish policy documents. There is however a recognition of the need for interdisciplinary research where researchers in both AHSS and STEM disciplines collaborate on themes of mutual benefit. The Irish Research Council (IRC) established a New Horizons Interdisciplinary Programme for AHSS-led research projects and partnership with the SFI Investigator Programme on STEM-led interdisciplinary research. The IRC has also sponsored events such as Inspirefest (2016), an international festival of technology, science, design and the arts. The term STEAM has emerged through stakeholders' efforts to secure extra investment in the arts. The National Campaign for the Arts (2015) in a prebudget submission referred specifically to STEAM when requesting more resources. The recently published government action plan for Education 2017 makes a commitment to identify and address skills gaps, ICT and STEM needs. It also gave a commitment to extend the Music Generation programme and to implement the Arts in Education Charter, subject to the drawdown of Dormant Account

Funding (DES 2017a, b). However, there is no reference to developing a STEAM approach.

Conclusion

Within Kingdon's (1995) Multiple Streams Framework, it is possible to identify patterns in the policy process with reference to STEM and STEAM. The exploration of policy directions facilitates a closer analysis of the challenges faced by STEAM initiatives securing mainstream policy support to become part of the policy window.

The need to enhance the supply of human capital and enlarge a high skill group capable in research and innovation was articulated by organisations such as the OECD and agencies within the EU as challenges to economic competitiveness. Interestingly these challenges were constructed in similar ways across countries worldwide. The range of policy solutions offered focussed on STEM as having a utility value that would in turn address these challenges, and the political context was mobilised to ensure further support. The convergence of the three streams privileged STEM making it very difficult to offer alternative policy solutions. The resultant policy practices witnessed increased funding and structural supports to STEM, where centrally driven and funded programmes were developed; curricular reform and new teaching standards were emphasised, and world-class university programmes were promoted, in addition to the recruitment of foreign science talent. STEM activities in schools, vocational and higher education were linked to industry, business and the professions. These policy practices facilitated a science model where impact was viewed in terms of the successful spin out of companies, patents and economic return. Consequently innovation became inextricably linked to the notion of technology transfer. There is a lack of agreed statistical definitions within countries of what constitute STEM study fields, STEM occupations and STEM sectors, and these gaps mean there is often a lack of adequate data to inform policy-making reliably. A broad set of areas were prioritised in policy development across countries. These included areas such as biotechnology and life sciences, ICT and nanotechnology. This suggests that governments were not prioritising according to national need but were imitating policy directions in other countries. The structured support for STEM in Europe was clearly reflected in the allocation of research funding under the multiannual research frameworks and in the more recent emphasis on global challenges and innovation. The reorganisation of European funding agencies reinforced the dominance of scientific dimensions of European research. Areas like the arts and humanities received less funding, and this further reinforced the perception that this type of research and activity was not as valid in the promotion of economic development and competitiveness.

The Multiple Streams Framework provides some very interesting insights into the Irish policy context. For over 40 years, it was recognised that science and industrial development were key challenges for the Irish economy. Beyond the publication of reports and the setting up of working groups to examine the issue, little else was achieved. The influence of European funding and the emergence of the 'Celtic Tiger' economy marked a shift in policy direction and resulted in more focussed efforts at promoting science as a route to economic success. The economic recession and onset of austerity focussed policy direction on repairing the economy. Various policy solutions emerged which included cuts to existing resources and ensuring that universities engaged in research that would have economic impact. The political conditions were conducive to supporting this approach, and the policy window emerged which provided STEM with support in structural and resource terms. This is reflected in the transfer of PRTLI from the Department of Education and Skills to the (then) Department of Enterprise, Trade and Innovation and the establishment of Science Foundation Ireland as the principal funding agency. The introduction of the Research Prioritisation Exercise (RPE) marked the end of what had been a strategy to build a broad base of expertise in favour of a more targeted approach. Research relevance was now defined principally in terms of job creation with an emphasis on science and technology. While there is no official STEM education policy, nevertheless a number of curricular initiatives in the STEM subjects have been introduced in the second-level education system.

The period of fiscal constraint in Ireland also led to a focus on the arts as having a role to play in economic revival. Traditionally the arts sector has been underfunded. Within the education system, the introduction of the Arts Charter and curricular initiatives emphasising creativity demonstrates a new emphasis on the arts. The Creative Ireland Programme also articulates the potential economic role for the arts. Very little emphasis has been placed on the humanities. Policy documents have highlighted the role that the humanities and social sciences can play in economic development, and while some important capital projects supporting the arts, humanities and social sciences have been realised, these areas have remained underfunded. The introduction of the Research Prioritisation with its emphasis on the economy and industrial development made it difficult to for AHSS to secure research funding. STEAM as a term has not featured prominently in official policy documents. The IRC and SFI emphasise interdisciplinarity as opposed to using the term STEAM. It is clear that the hegemony of STEM retains its prominence.

More recently the challenge of innovation and creativity in economic competitiveness has been recognised. The realisation that there is a need to link the arts with the sciences as key education and workforce policy concerns has resulted in more discussion within the policy context. However there remains ambiguity concerning the terms innovation and creativity. For those promoting STEAM, it is clear that they face many challenges in establishing this approach as a viable policy alternative. It is not clear that the political stream will support such an approach over the longer term, whether the educational community will embrace it and whether industry will champion STEAM in the same way that STEM has been promoted.

References

Ball P (2015) Opinion: 'oh the humanities'. Chemistry World. Royal Society of Chemistry. https://www.chemistryworld.com/opinion/oh-the-humanities/8460.article

Benneworth P, Gulbrandsen M, Hazelkorn E (2016) The impact and future of Arts and Humanities Research. Palgrave Macmillan, Basingstoke

Benson C (1979) The place of the arts in Irish education: report of the working party appointed by the Arts Council. Arts Council of Ireland, Dublin

Chow A (2014) Understanding policy change: multiple streams and National Education Curriculum Policy in Hong Kong. J Public Adm Gov 4(2):49–64

Clancy P (2015) Irish higher education. A comparative perspective. Institute of Public Administration, Dublin

Clarke M, Kenny A, Loxley A (2015) Creating a supportive working environment for academics in higher education: country report Ireland. Educational International, Irish Federation of University Teachers and Teachers Union of Ireland, Brussels/Dublin

Coolahan J (2016) Arts in education charter from 2003 to 2016 –where to now? Education Training Boards of Ireland, Issue 3, pp 5–8

Coopers, Lybrand (1994) The employment and economic significance of the cultural industries in Ireland. Corporate Finance Service, Dublin

de la Garza A, Travis C (2017) The STEAM revolution bridging the divide: transdisciplinary approaches to arts, humanities, science, technology, engineering and mathematics studies. Springer, Cham

Department of Arts, Heritage and the Gaeltacht (2015) Value for Money and Policy Review of the Arts Council

Department of Education and Skills (2017a) Proposed Exchequer – Employer investment mechanism for higher education and future education and training. Consultation Paper. https://www.education.ie/en/Publications/Education Reports/pub_ed_proposed_exchequer_employer_investment_higher_further_training_2017.pdf

Department of Education and Skills (2017b). Action Plan for Education 2017

Department of Expenditure and Public Sector Reform (2014) Comprehensive expenditure review. http://per.gov.ie/wp-content/uploads/CER-Estimates-Final21.pdf

Dill DD, van Vught FA (2009) National Innovation and the Academic Research Enterprise. John Hopkins University Press, Baltimore

Eivers E, Close S, Shiel G, Millar D, Clerkin A, Gilleece L (2010) The 2009 national assessments of mathematics and English reading. Educational Research Centre, Dublin

European Commission (2015a) Science education for responsible citizenship: report to the European Commission of the expert group on science education. Directorate-General for Research and Innovation science with and for society: EUR 26893 EN

European Commission (2015b) Does Europe need more STEM graduates. Final report. Danish Technological Institute, Danish, pp 1–68

European Commission (2017) Communication from the Commission to the European Parliament, the Council, the European Economic and Social Committee and the Committee of the Regions on a renewed EU agenda for higher education. Brussels, 30.5.2017 COM(2017) 247 final, pp 1–12

Fitzgibbon M (2011) Forty years of Irish industrial and science policies, 1960-2010: a critical analysis. Administration 59(3):101–141

Galligan A (2014) U.S. Workforce policy builds up 'STEAM'. Cult Work! 18(2):1–6

Gulbrandsen M, Aastad S (2015) Is innovation a useful concept for arts and humanities research? Arts and Hum High Educ 14:19–24

Harpur J (2010) Innovation, profit and the common good in higher Education. The New Alchemy. Palgrave MacMillan, London

Hazelkorn E (2014) Making an impact: new directions for arts and humanities research. Arts and Hum High Educ 14(1):25–44

Hazelkorn E, Gibson A (2016) Ireland: valuing the arts and humanities in a time of crisis and beyond. In: Benneworth P, Gulbrandsen M, Hazelkorn E (eds) The impact and future of arts and humanities research. Palgrave Macmillan, Basingstoke

Indecon International Economic Consultants (2011) Assessment of economic impact of the arts in Ireland: an update report. The Arts Council, Dublin. Available at: http://www.artscouncil.ie/Publications/Indecon_Update_Report_n.pdf

Kingdon J (1995) Agendas, alternatives and public policies, 2nd edn. Longman, New York

Land M (2013) Full STEAM ahead: the benefits of integrating the arts into STEM. Proc Comput Sci 20:547–552

Maeda J (2013) STEM + Art = STEAM. STEAM J, 1(1)

Marginson S, Tytler R, Freeman B, Roberts K (2013) STEM: country comparisons: international comparisons of science, technology, engineering and mathematics (STEM) education. Final report. Australian Council of Learned Academies, Melbourne

May G (2015) Stem not STEAM, Opinion. Inside Higher Education. https://www.insidehighered.com/views/2015/03/30/essay-criticizes-idea-adding-arts-push-stem-education

Nutley SM, Walter I, Davies HTO (2007) Using evidence: how research can inform public services. The Policy Press, Bristol

Organisation for Economic Co-operation and Development (2011) Education for innovation: the role of arts and STEM education Workshop Summary Report. EDU/CERI/CD(2011)8

Pollitt C (2008) Time, policy, management: governing with the past. OUP, Oxford

Seery A (2008) Slavoj Zizek's dialectics of ideology and the discourses of Irish education. Ir Educ Stud 27:133–146

Smyth E (2016) Arts and cultural participation among children and young people: insights from the growing up in Ireland study. Arts Council of Ireland and Economic, Social and Research Institute, Dublin

STEM Education Review Group (2016) STEM education in the Irish school system. A Report on Science, Technology, Engineering and Mathematics (STEM) education analysis and recommendations, pp 1–72

van den Doel HW (2013) The future of the Social Sciences and Humanities in Europe: collected LERU papers on the SSH research agenda. League of European Research Universities (LERU)

West S (2013) Humanities research leadership in Europe: stimulus paper. Leadership Foundation for Higher Education, London, pp 1–36

Zahariadis N (2003) Ambiguity and choice un public policy: political decision making in modern democracies. Georgetown University Press, Washington, DC

Marie Clarke is Dean of Undergraduate Studies in University College Dublin and is an Associate Professor in the School of Education. She researches and is published in higher education, education policy and history of education.

Chapter 16
STEAM: Powering the Digital Revolution

Leonard Hobbs

Abstract As the digital revolution continues to drive the world's evolution in the 'Information Age', it is generally accepted that science, technology, engineering and mathematics (STEM) skills will be crucial for humanity to be able to tackle many of the grand challenges that lie ahead and for enabling the kind of innovation the world's economies require to grow. However it is now being questioned as to whether this somewhat singular focus on STEM skills alone, at the expense of a diminishing focus on the more creatively centred arts subjects, will deliver the many needs of a resource-stretched world. This chapter argues that the time is right to promote the concept of STEAM where the more traditional arts and humanities subjects are blended with the STEM. It reflects on the lessons from history by looking at some of the world's greatest innovators who were practitioners of STEAM and whose many breakthroughs were enabled by a combination of the creative and innovative processes. It goes on to describe how diversity in the workplace, which is an accepted catalyst for business improvements, can be optimised by the application of a STEAM agenda by enterprises.

Keywords CoderDojo · Skills · Workforce · The Ford Motor Company · Nokia Bell Labs · Creativity · Standardisation · Lean methodologies · Diversity · Inclusion · Education

From STEM to STEAM

For the past number of years, the focus on education in the so-called STEM subjects has been receiving attention on a global scale as nations look to deliver the kinds of skills that the workforce of the future will require. Former US Secretary of Education Richard Riley asserted (Jones et al. 2004) that 'the top 10 in-demand jobs in the future don't exist today will use technologies that haven't been invented and solve

L. Hobbs (✉)
Research and Innovation at Trinity College Dublin, Dublin, Ireland
e-mail: leonard@leonardhobbs.com

© Springer International Publishing AG, part of Springer Nature 2019

A. de la Garza, C. Travis (eds.), *The STEAM Revolution*,
https://doi.org/10.1007/978-3-319-89818-6_16

problems we don't even know are problems yet.' The European Parliament (Europarl.europa.eu 2015) has estimated that there will be 7 million job openings in STEM positions in Europe in the next 10 years. In China, it is estimated that the spending on STEM education will drive a $15 billion private education industry by the end of the decade as parents are looking to get their children skilled in coding and robotics. In Ireland, what began as a coding club, called CoderDojo (CoderDojo.com 2017), for 7–17-year-olds in 2011 has now grown into an international community with over 1000 clubs in 63 countries teaching children and teenagers how to build websites, create a game or work with technology. However, in a world which is becoming progressively 'more flat' (Friedman 2005) as the digital revolution itself delivers global access to knowledge and capabilities, enterprises are finding it more challenging to differentiate themselves. Businesses today need to create breakthrough products and services so that they can stay ahead of their competition and avoid a future which is only based on improving productivity and reducing costs.

The industrial revolution of the 1800s was largely enabled by the STEAM engine. As the digital revolution takes hold in today's world, we once again encounter 'STEAM' as an enabler, only on this occasion, it is the integration of the creative arts with the STEM subjects which is being referred to. In the USA, where today's focus on STEM skills had first emerged, there is a growing concern (Humanitiescommission.org 2013) that they are now being left behind as other economies such as China (Chen 2016) and Europe are looking to broaden their education to include the arts. These latter regions recognise the creative subjectivity of the arts and social sciences as being an inspiration for further innovation when integrated with the traditional STEM.

Private enterprise is starting to wake up to the opportunity which STEAM brings as they look to hire artists and designers (Beal 2013) who can bring a unique view on problem solving and user experiences to the workplace. There is an emerging understanding that the skills of the workforce of the future will depend as much on curiosity, creation and design as technical aptitude. The Ford Motor Company sponsor programmes at high school which seek to 'fire up the artistic right brain along with the logical left brain to give kids a broader view that inspires even more inventive and innovative thinking' (Campaign-social.ford.com 2017). The Sesame Street group have introduced the STEAM concept in their 43rd season with shows such as 'Elmo the Musical' (Wong 2013) combining music and maths for the young viewers. At Nokia Bell Labs, they have revived their Experiments in Art and Technology programme (EAT 1970), from the 1970s, to investigate the intersection of the arts and technology. They actively collaborate with musicians and arts and fashion designers to create new experiences at this intersection and to explore how communication technology will continue to drive the human experience. According to CTO Marcus Weldon, they 'will create the future of connected wearables that elicit, transmit and receive multidimensional sensory experiences allowing us to explore the future of human communications'. In Ireland, which is a country known for its creative minds in the arts world, there is an emerging interest amongst educators and enterprise into how the STEAM concept might give Ireland an 'edge'. This has been explored at the iconic 'Other Voices' musical festival (Irelands Edge 2016) in the remote town of Dingle in County Kerry where companies such as Intel, PWC and Microsoft have engaged with musicians and performers in discussing the relationship between the arts and technology.

Lessons from History

The need for STEM-related skills is driven by the need for innovation that can translate scientific ideas and knowhow into practical technologies that have socio-economic impacts. However, some of the world's more innovative people have also been hugely creative, as indicated by their abilities as artists, poets and musicians. Indeed, many were early practitioners of the STEAM concept.

Leonardo Da Vinci (1452–1519) was probably the most proficient at combining the STEAM disciplines. Although he is best known for his paintings, such as the *Mona Lisa* and the *Last Supper*, he was also an inventor, mathematician, musician and writer. He considered that a person's sight was the most important sense of all, as it enabled him/her to observe the world in all its glory. He became an expert in the anatomy of the human body and combined this scientific knowledge with his artistic talent to produce many detailed drawings. One of the most famous of these was the 'Vitruvian Man' in which he combined mathematics and art to explore the theories of the Roman architect Vitruvius on the proportions of man. In observing bird flight, he also used his artistic talent to conceive flying machines that resembled modern-day hang gliders and helicopters.

Over 100 years later, Galileo (1564–1642) combined natural philosophy, astronomy and mathematics to advance the scientific fields of motion, astronomy and materials. Most notably from a STEAM perspective, he applied mathematical methodologies to the writing of articles on nature, thus transforming what was a subjective approach into a more objective experimental method for recording discoveries in nature.

Samuel Morse (1791–1872) who was the coinventor of the electric telegraph and developed the language which was used to translate and transmit messages called Morse code started out in life as a portrait painter, and indeed some of his works are considered to be some of the finest portraits ever completed by an American artist. However, it was his early interest in the then emerging field of electricity and magnetism which eventually drove his ambition in a more technical direction after an encounter on an art expedition to the UK. He would go on and lay the world's first telegraph line between Baltimore and Washington in 1844, and over 20 years later, he also helped with the first transatlantic cable which linked New York to London, via Valentia Island in Co Kerry Ireland, thus beginning the era of global communications.

George Boole (1815–1864) was the first professor of mathematics at University College Cork. Boole proposed that logic should be aligned to mathematics which he detailed in his 1854 book *An Investigation into the Laws of Thought on Which Are Founded the Mathematical Theories of Logic and Probabilities*. Almost 100 years later, a young engineering master student at the Massachusetts Institute of

Technology, called Claude Shannon, came across Boole's publication while studying symbolic logic in an undergraduate philosophy course and later combined the concepts to invent a new way of designing electrical circuits. This new method would rely on the Boolean principles of true and false, which when expressed as 1's and 0's (knows as 'bits') could be used to design 'digital' circuits. This methodology would go on to change the world as it enabled the growth of the telecommunications industry and subsequently the computing industry which have themselves now evolved into the pervasive Internet industry. Today these digital circuits are used to connect billions of people and machines in seconds across the world using the electronic 'bits' of information, all enabled from the merger of logic, mathematics and engineering.

Albert Einstein (1879–1955) who is probably the most famous scientist who ever lived had an avid love of classical music and was a violinist. He was once quoted as saying that 'life without playing music is inconceivable for me' and was deeply moved by the music of Mozart which he said was so 'pure and beautiful that I see it as a reflection of the inner beauty of the universe itself'. He saw similarities with his own creative processes as he first imagined and then translated some of the mysteries in the physical world.

Standardisation, the Creativity Killer

As today's modern industries look to be more competitive amidst fierce global competition, many companies' innovative energies are directed at greater efficiencies and productivity improvements. Standardisation techniques are widely implemented across global company networks, which result in greater efficiencies by copying methodologies, removing waste and improving quality. The pursuit of ISO9000 standards (Iso.org 2017) became very popular in the 1990s as the focus was on improving customer quality by addressing standard procedures in companies via lengthy and detailed documentation.

The competitive drive has also led to reduced activity in the more creative disciplines such as in research and development. Today the trend is for large companies to reduce their spending on research or outsource the activity to large research centres. Also, more companies are looking to grow by means of an acquisition path in which the creativity of smaller companies is assimilated by the larger organisation. In 2012 tech giant CISCO (Scott 2012) spent billions of dollars acquiring 9 companies, and in 2015, the amount of acquisitions in the pharmaceutical industry grew by 94% from the previous year (Fisher 2015). Research organisations such as Belgium's IMEC have benefited from this transition as they attract more companies to do their research at their facility and have grown to be a world-leading semiconductor research centre, with 3500 scientists, in a country which doesn't even have a semiconductor business. This relentless cost improvement drive is often enabled by the application of 'lean' methodologies, which has its origins in the 'just in time' manufacturing at the Ford Motor Company. It was further developed some time later by the Toyota Motor Company who looked to improve upon US manufacturing capability in post-war Japan.

The lean methodology is comprised of five principles (Flinchbaugh and Carlino 2006) which include directly observing work activities, systematically eliminating waste, standardising work practices, solving problems in a systematic way and creating a learning organisation. In recent times the system has migrated beyond manufacturing companies into service-orientated companies with Ryan Air being a prime example. In the early 1990s, they turned a struggling small airline (Ruffa 2008) into one of the most successful airlines in the world by continuously and consistently applying waste elimination improvements, ranging from the removal of magazine holders on the back of seats to deploying the first online booking system. While the methodology does call for the creation of a 'learning organisation' in which its 'processes are considered as laboratories and its employees are its scientists', this creative aspect is often squeezed out in the pursuit of quick productivity gains in the fast-changing globally connected world of today.

However, one does not need to only look for companies which have implemented lean principles. The power of standardisation has been implemented across many industries, driven by the increasing effectiveness of large global players who optimise their supply chains across the globe, focusing on cost reduction by deploying standard procedures and productivity enhancing tools in lower cost economies. The global harmonisation has led to many of our day-to-day products and services being the same, with global brands dominating the high street, online services with the same offering, no matter where you access them across the world, in such a way that Coca-Cola, for instance, must now share the stage as only one of many global brands in the food and beverages sector. In the car industry, there is little difference between different models, apart from cosmetic, with most cars performing at the same level. Indeed the car industry today is now relying more and more on the innovative capability of electronics to provide a more differentiable user experience, with companies such as Germany's Audi stating that 80% of their innovation now comes from electronics and not their core competency, mechanics (Hellenthal 2017).

The lead motor companies are now reaching out to the more innovative information and communication technology (ICT) sector as the latter sector looks to add more user experiences to the motoring journey and improve safety features. In the meantime, the larger and more creative ICT companies, such as Apple and Google, are approaching in the other direction with the development of autonomous cars. Google (Davies 2016) produced the first self-driving car in 2015. The winning combination at the interface of these two mega business sectors will no doubt be those companies which converge the best of the creative and design competencies, with the innovative and implementation capabilities in producing the breakthrough products and user experiences. Once the technology does appear on our streets in the decade ahead, companies will need to look at their creative capabilities in producing new user experiences in the car as the 'driver experience' will become a distant memory.

Diversity and Inclusion Driven by STEAM

In its 2015 paper 'Diversity Matters' (Hunt et al. 2015), McKinsey reported that diversity is a competitive differentiator. In US companies, once women in the workforce constitute at least 22% of the senior executive team, a correlation is observed of 0.3% growth in earnings for every further 10% increase. There was an even greater return for improvements in ethnic diversity with a 0.8% improvement being observed for a 10% increase in this diversity category. The correlations are even stronger for UK companies with a 3.5% improvement in earnings for every 10% increase in gender diversity.

Although gender diversity has received much attention in recent years, the topic is now broadening out to a wider treatment of the issue as it is evident that companies can benefit from getting the right blend of all diversity categories. Also, as companies are discovering the benefits of becoming more open in their innovation endeavours, particularly in the cross-sectoral and cross-disciplinary collaboration with customers, suppliers and partners (Oitcinterfor.org 2014), the need for greater diversity of thought and capability is key. At Intel, they believe that diversity and inclusion are amongst the 'most important forces driving that evolution and reinvention' (Brown 2017), while at Mars UK (Dawson 2015) even though they have over 50% of female managers, they believe that it is a 'Trojan horse talking about gender because it's about a lot more than being male or female. It's also about inclusivity and diversity and getting a broader group'.

Diversity in its broadest sense is often described in categories, such as gender, sexual orientation, disability, race, religion or age. However, diversity can also describe personal characteristics and attributes such as those uncovered in the application of emotional intelligence (Goleman 1996). Personal analysis tools are often used to reveal this level of diversity. Myers Briggs (Myersbriggs.org 2017) identifies 16 distinctive personality types such as the 'commander' who is bold, imaginative and strong-willed or the 'mediator' who is poetic, kind and eager to help a good cause. Marcus Buckingham's work (2001) focuses more on identifying strengths and describes roles such as the 'pioneer' who sees the world as a friendly place and whose strength comes from optimism in the face of uncertainty, or the 'stimulator' who feels responsible for others and looks to elevate them. Although popular as management tools, these diversity-uncovering programmes are of little use in terms of people's effectiveness in the workplace unless they are also accompanied by a genuine willingness on behalf of the employer to create the time and space where employees can fully express their true selves. In her 2011 book *Flourishing* (Gaffney 2011), Dr. Maureen Gaffney described a number of elements which would enable one to be 'at your best'. These included the need for a distinct challenge, being connected internally and externally, having a high degree of autonomy and also using your valued competencies.

In the 'lean'-driven world in which we live, the drive for standardisation as a method to improve costs and productivity has a knock-on effect in that people's broad diversity is 'leaned' out as there is little room for what is different. The modern office space is a good indication of such a phenomenon as spaces are typically shared and space-efficient open plans afford little room for personalisation. Employees are all given the same smart phone and company calendars, driven by the need to be globally connected, leaving little space for reflection. If people are to 'be at their best' and enabled to use their full set of competencies, then the starting point needs to be a realisation of the fact that no two employees are the same and in fact all are 'diverse'. It is the company's ability to unpack its wealth of diverse capabilities which will not only raise all diversity boats, including categories such as gender and race, but ultimately deliver greater overall performance.

The application of a culture of STEAM in the workplace would enable employees to bring their 'complete selves' to the work place, which would lead to be a more inclusive, contented and collaborative organisation by colliding the creative with the practical, the musician with the engineer, the artist with the scientist, producing new ideas, projects and breakthrough solutions. Efforts would need to be made to create an inclusive workplace where employees were free to exploit their diverse personalities and associated interests. Starting with the workplace design, companies should look to incorporate art work into the spaces which would provoke conversations, as opposed to today's 'vanilla' designs with pastel shades. Companies should implement 'diversity programmes' which look to identify the differing personalities and should promote opportunities for employees with like interests to collaborate. Organisational structures and teams should be constructed with diversity in mind so that there is a good sprinkling of different types up and across the organisation. Analysis should be performed regularly to understand how the different diverse types are distributed across the organisation and care should be taken to avoid the dominance of one type to avoid 'groupthink'. Company calendars should include time to reflect on the problems and opportunities which the company faces, and involve alternative thinkers, such as artists, poets or historians, to address the meetings with a view to promoting an alternative way of thinking about the challenge. Companies should develop hiring tools which will explore the diversity of the interviewee with the intent of adding more 'colour' to their workforce. As well as organising programmes which give the space for employees to express the 'A' part of skill set, companies should also look to hire people who are predominantly 'A' types, such as artists, social scientists and humanities students. Artists can bring a level of creativity to an organisation which can enable new product designs, social scientists understand how new services can be best deployed in the world, historians can provide a reflection in the transformational nature of previous technologies and poets interpret the world through and in a different dimension. A more recent example where a company took advantage of a more lateral approach is the Apple computer company who have combined beautiful design with innovative use of materials and services to build one of the world's largest companies with net worth of $200 billion in yearly revenue. In creating new products, the company relies on its creative capabilities and looks to 'start again by disconnecting from the past' (Ive 2012).

In the increasingly busy lives we now lead, people are in constant connection and the expectation for an almost instantaneous response is very real. Employees find it more and more difficult to strike the 'work-life' balance as work time seems to increasingly encroach on the away-from-work time. In the application of such, these STEAM-orientated programmes would also go some way to improve the 'work-life' balance equation as employees would have the opportunity to import some of their 'life skills' to the work place. In so doing they can improve the balance by bringing their life to work as opposed to always having work encroach on their life.

A report from the Center for Talent Innovation (Mundy 2017) found that in the USA, when women drop out of technology companies (and women in USA leave at twice the rate of men), it is usually not for family reasons. They report a lack of access to key creative roles is one of the key reasons.

STEAM and Education

As described previously, some of the world's greatest innovators have also been hugely creative, and they effectively combined their complete skill set in achieving paradigm-shifting breakthroughs. In today's world, our education system begins with a broad base when young children are exposed to maths, music, nature, language and science. As they progress through the system, their exposure gets narrower so that when they reach university level, they are now following a narrow path which narrows even further when they enter the workforce and look to drive their careers by being experts in a particular area. Our education system is designed to produce 'tall, skinny' minds with any 'T shaping' only occurring as a result of informal training in the pursuit of personal interests. Our workforce design copper fastens this by pigeon holing employees in specific roles where there is little time for creativity and lateral thinking.

In a similar way to businesses, the education system has also been shaped by the need to standardisation so that many more students can be pumped through the system in an efficient manner. Innovation within the system is extremely slow and cumbersome such that new essential skills such as 'coding' are slow to appear on curricula. In recent times the emphasis on the STEM subjects has taken hold (MacCraith 2016) with incentives being offered such as bonus points for those students who take honours maths at second level (IUA.ie 2012). However, this rush to STEM is further isolating the creative disciplines, and the wonderful discovery-based education of the primary level is left behind. This policy is continued into the third level, where large government-funded STEM R&D programmes are prevalent, and there is little scope for the more creative pursuits. A STEAM education policy could look to broaden the educational base from first to third level and would deliver both creative and innovative graduates to the workforce.

A Final Thought

Oscar Wilde recommended to 'be yourself, everyone else is already taken'. As most countries around the world look to produce more people with STEM skills so as to fuel the continued progress of the 'digital revolution', those countries and indeed companies, who embrace the STEAM concept, will have a greater chance to differentiate themselves by unlocking the true potential of their people, embracing their unique diversity and unleashing their creative abilities in pursuit of continued innovation.

References

Beal S (2013) Turn STEM to STEAM: why science needs the arts. S Beal, Huffington Post blog. Available at: http://wwwhuffingtonpostcom/stephen-beal/turn-stem-to-steam_b_3424356html November 2011
Brown D (2017) Reflections on intel's diversity & inclusion journey: 2016 diversity & inclusion annual report. Available at: www.intel.comt. Danielle Brown, Chief Diversity & Inclusion Officer Feb 201
Buckingham M, Clifton D (2001) Now discover your strengths. Gallup Press, Buckingham
Campaign-social.ford.com (2017) Mission statement. Available at: https://campaign-social.ford.com/content/campaign/ford-steam/mission.html
Chen LY (2016) China's Tiger moms are spending big on tech classes for their kids. Bloomberg publication available at: https://wwwbloombergcom/news/features/2016-12-20 – LY Chen, Bloombergcom December 2016
CoderDojo.com (2017) CoderDojo movement. Available at: https://coderdojo.com/movement/
Davies A (2016) Google's self-driving self driving car company is finally here. Article in Wired.com. Available at: www.wired.com/2016/12/google-self-driving-car-waymo/Waymo.co
Dawson F (2015) Irish woman finds sweet spot at the heart of Mars. Irish Times article available at: wwwirishtimescom. Oct 23rd 2015 with Fiona Dawson of Mars, Feb 2015
EAT (1970) Available at: https://www.bell-labs.com/explore/experiments-art-and-technology/
Europarl.europa.eu (2015) Encouraging STEM studies for the European Market. Available at: www.europarl.europa.eu – EU parliament document 2015
Fisher N (2015) Are M&A replacing R&D in pharma? Forbes Magazine N Fisher. Available at: wwwforbescom/sites/nicolefisher/2015/04/22/. Forbes April 2015
Flinchbaugh J, Carlino A (2006) The Hitchhiker's guide to lean. J Flinchbaugh, A Carlino, Society of Manufacturing Engineers, 2006
Friedman T (2005) The world is flat: a brief history of the twenty-first century. Farrar, Straus and Giroux T Friedman 2005
Gaffney M (2011) Flourishing. Maureen Gaffney, 2011 Penguin Ireland
Goleman D (1996) Emotional intelligence. Bloomsbury Publishing – Daniel Golema, Londres
Hellenthal B (2017) Cross-industry collaboration networks accelerate innovations. B Hellenthal. Available at: http://www.semi.org/eu/iss-europe-2017
Humanitiescommission.org (2013) The heart of the matter, the American society of the arts and sciences. Available at: http://www.humanitiescommission.org/_pdf/hss_report.pdf
Hunt V, Layton D, Prince S (2015) Diversity matters. McKinsey & Company
Ireland's Edge at Other Voices (2016) Available at: http://www.othervoices.ie/content/irelands-edge
Iso.org (2017) ISO 9000 – quality management. Available at: www.iso.org/iso-9001-quality-management.html

IUA.ie (2012) Bonus points for maths. Available at: http://www.iua.ie/students-graduates/admissions/bonus-points-for-maths/
Ive J (2012) The new soul of apple video. Available at: http://www.youtube.com/watch ?v=_ynXKHC9Wo4. At SEMI ISS Munich March 2
Jones R, Scanland K, Gu S (2004) The jobs revolution: changing how America works. Copywriters Inc
MacCraith B (2016) A report on science, technology, engineering and mathematics. A report atics Education to the Minister for Education and Skills in Ireland. Available at: www.education.ie/en/Publications/Education-Reports. Nov 2016
Mundy L (2017) Why is Silicon Valley so awful to women. The Atlantic. Available at: www.the-atlantic.com/magazine/archive/2017/04. The Atlantic, April 201
Myersbriggs.org (2017) MTBI basics. Available at: www.myersbriggs.org/my-mbti-personality-type/mbti-basics/
Oitcinterfor.org (2014) The future of work. Jobs and skills in 2030. UK Commission for Employment and Skills, February 2014. Available at: https://www.oitcinterfor.org
Ruffa S (2008) Going lean: *how the best companies apply lean manufacturing.* Amacon S Ruffa, 2008 business and economics
Scott J (2012) Why CISCO choses acquisitions over R&D. Article in computer weekly. Available at: www.ComputerWeekly.com
Wong S (2013) Create your own stories with Elmo. Available at: www.sesameworkshop.org/press-releasesSocial.ford.com

Part VI
Contra STEAM

Chapter 17
Contra-STEAM: On the Value of the Autonomy of Art

Robert F. Arnold

All art is quite useless – Oscar Wilde

Abstract While recognizing the educational and ultimately societal values of STEM initiatives to confront gender imbalance in science and technical fields early on, as well as the virtues of interdisciplinary and collaborative problem-solving associated with STEAM, this chapter will tease out and call into question some of the assumptions and possible impacts of including art in that mix of disciplines, from the standpoint of a longstanding split in the theorization of art between utilitarian and non-utilitarian conceptions of aesthetic practice. While incorporating art into STEM has many potential benefits, does it ultimately undermine the societal value of art as an independent, autonomous sphere, that sits outside of the practical utilitarian framework of STEM? Does it sacrifice art's potential to serve as a critical mirror of society, or as an escape from the practical, to put it in service to the ideology of "creative problem-solving?"

Keywords Art and ideology · Art as a form of knowledge and knowing · Autonomy of art · Neo-Kantian aesthetics

When asked to contribute an essay on STEAM from an artist's perspective, I hesitated, lacking more than a passing understanding of the idea and also because of how ambivalent I felt toward the subject. What I offer is a personal take on the question that possibly says less about STEAM than about my own attempt to understand that ambivalence.

My understanding is that STEAM grew out of STEM, a laudable initiative to confront a real problem concerning the underrepresentation of women in the

R. F. Arnold (✉)
Montana State University, Bozeman, MT, USA
e-mail: rfarnold@montana.edu

© Springer International Publishing AG, part of Springer Nature 2019
A. de la Garza, C. Travis (eds.), *The STEAM Revolution*,
https://doi.org/10.1007/978-3-319-89818-6_17

science, technology, engineering, and math fields, as well as the need to address a shortage and lack of preparation of students entering those fields in general. Although these are not my fields, I believe that I am sensitive to issues of gender through my study of film and also as the father of a daughter now in middle school. I was teaching at a college in Boston at the time of Harvard President Larry Summer's infamous remarks about women in science, attributing the gender imbalance to "innate" differences in aptitude. Everything I heard about STEM made sense, but because these are not my fields, I had no direct connection to any STEM activities.

However, it wasn't long after public universities, including the state university where I teach now, embraced STEM concepts, an outcry erupted to include art in the mix, and thus was born STEAM. My sense of any real problem that STEAM initiatives were intended to address started to diminish. At least at my university, the problem became the perennial problem of resource allocation. If new resources were to be devoted to STEM, then the arts had to find a way to claim their slice of the pie. A former Dean of my arts college saw it as her mission to put the "A" into STEM and set her focus on presenting our college as the creative spark that would ignite the interdisciplinary possibilities of STEAM. It should be noted that this Dean had little personal experience in the creative arts herself, and STEAM seemed like an idea she could understand and "sell" in order to leverage the college's – and her own – position.

There seems to be a recurring pattern with universities attempting to confront their compartmentalized organization. Interdisciplinarity is a virtue that represents the ideal of a university as a community of intellectuals who share, learn, and discover together. However, the reality is by and large disappointing, increasingly so as budgets decline and competition for resources grows, along with the emphasis on "the bottom line" measured by whatever "metrics" are in vogue, pushing aside the more laudatory but perhaps less practical goals that now seem quaint, such as having a "well-rounded" liberal arts education.

I am a product of such a liberal arts education perhaps at its best, attending a small Midwestern college that was a pioneer of interdisciplinary education in the 1970s. I remain a firm believer in liberal arts ideal that the ultimate goal of a college education is to produce a well-rounded, critically thinking (i.e., "skeptical"), self-aware human being. Call me old-fashioned (or, if you wish, a child of former middle-class privilege). Naturally, therefore, I believe strongly that scientists should also study art and artists should also study science and so on. However, it seems to me that as this ideal began to fade as education budgets were slashed and tuition costs rose and practical careerist goals took precedence, colleges and universities have been locked in a cycle of trying to reclaim what has been lost in ever new-fangled ways.

As I understand it, the goal of STEAM was to ensure that the arts would not be forgotten as our focus shifted toward STEM. This reflects that lost sense of an integrated (not necessarily interdisciplinary) holistic liberal education. Frankly, it seems something of a last gasp to me. As an artist (filmmaker) and as an academic who teaches in an arts college, I'm all for including the arts in the conversation and the distribution of resources. But with each new iteration of integration, it seems that things are often lost in translation, and it is in trying to understand what the arts bring to STEM to make STEAM that I begin to have doubts.

At least here at my university, the mantra that is endlessly repeated as an argument for placing the arts in the center of the STEAM mix is the phrase, "creative problem solving." My goal in this essay is to think through this slogan that I believe serves to reduce both STEM and ART to, if not opposing then at least separate domains as binary opposites. However, as I will argue, art and science and technology and perhaps to a somewhat lesser degree engineering and math have always been important, vital, aspects of art on many different levels. They have always been interconnected. And by the same token, the areas that make up STEM have never lacked creativity. Art cannot claim a monopoly over creativity as a force that drives all human activity. The drive to discover, to know, to express, and to solve problems, technical or otherwise, is the essence of creativity. Art is not a "special sauce" that can spice up otherwise creatively void disciplines. And finally, despite the dynamic and ever-changing tensions between university departments as separate domains, or "silos" as we now tend to call them, and efforts to break down the barriers between disciplines to reclaim a lost semblance of integration, the arts and sciences are not separate, never were nor ever will be. Departments may come and go and reconfigure, but humans seek knowledge, create, and express themselves in overlapping ways.

The danger of STEAM as I see it is the tendency to reduce art to this express function of providing the creative juice to real-world, practical problems. I can't say but perhaps theoretical mathematicians also fear the impending demands of practicality that can be quantified by university administrations. I also fear, coming from a highly technical creative field (film) that requires expensive resources, that our needs will be understood differently as art becomes an adjunct to the "real" technical fields. My goal here is to challenge some of these underlying assumptions that lead to these concerns.

I am by no means the first or only person to express concerns about STEAM. Some are concerned that including art will dilute the important goals of STEM (May 2015). My focus is different, concerned with what the inclusion of art might mean in relation to the autonomy of art as a necessary condition of its unique potential. I assume that STEM has an abundance of proponents that will safeguard its mission. I write in defense of the autonomous mission of art that I fear will be undermined by its inclusion in STEAM.

To some extent what is at stake is what one means by "art" and also what art "means" in terms of its social function(s). Suffice it to say that it is well beyond the scope of this paper, or my comprehension, to address these questions in their totality. For the sake of simplicity and brevity, I will equate "art" with the visual arts. This is not to say that the performing arts, architecture, literature, etc. aren't relevant. It is simply easier for me to discuss the visual arts because of my training and artistic practice in that area. I believe, however, that many of the points I hope to make might equally apply to the other arts.

As a historically determined "way of seeing" – borrowing John Berger's phrase from his 1973 BBC television series, "Ways of Seeing," and the book of the same title – art has always had close ties with the development of the evolving technologies and other social and material practices by which we understand our world. It is one of the ways that we express our understanding of that world to ourselves. The intersection of art and science, as well as technology and engineering, etc., in the Renaissance is very well known, personified by the polymath Leonardo da Vinci who, among many other things, painted religious and secular paintings and designed weapons of war. The development of optics used by Galileo's telescopes to see further into the heavens challenged ecclesiastical dogma and inaugurated the principles of the scientific method. At the same time, monocular perspective was introduced into visual representation achieving a new level of objective realism while at the same time inscribing the viewer's individual perspective as the privileged subject of visual representation, as the visual corollary of the age of humanism. This new "way of seeing," as Berger describes it, cut across all areas of human endeavor from science to cartography and exploration. The example of Italian Renaissance painting illustrates how the art of painting not only depicts an array of subject matter but also an understanding of "seeing" or perceiving itself. The mathematical geometry of linear perspective bespoke of an ordered universe that Newton would later describe scientifically. To illustrate the point at the risk of doing injustice to art history, one can see similar correlations between *en plein air* impressionist painting and concurrent theories of light as material, and later, between analytical cubism, combining multiple perspectives in a single picture, with the development of the theory of relativity, just to cite two examples.

In addition to this correlation between painting "style" and science, the tools of art making, always technological, also develop as other forms of technology develop and at times contribute to that development. Fine hair paintbrushes and oil paint replaced pigmented plaster and mosaic, permitting not only finer detail enabling individualized portraiture, but also a clearer indication of the unique "hand" of the artist. As industrialism displaced handwork, replacing the unique original handmade object, its uniqueness bearing the marks of its individual creator, with mass-produced identical components, painting as a means to visualize the world was displaced by the advent of photography and eventually film. As Walter Benjamin problematized, these arts, rooted in industrial mechanical and photochemical means and processes, designed for almost infinite reproducibility, lack an original object (Benjamin 1969 [1936]). As many scholars have pointed out, the industrial world is increasingly experienced through mass-produced *media*. The connection between photography representation and reality is both more objective, being less subject to human subjectivity and more based on optical and mechanical systems, and simultaneously more remote as the proliferating reproduction supplants the real thing. That tenuous connection between *analog* media and the word they measure and reflect is further strained by the more recent rapid development of digital visual media, recently achieving simulacra almost indistinguishable from reality and nearly infinitely plastic in terms of mutability.

The give and take between art and science and technology is constant. Brunelleschi's practical demonstration of the principles of linear visual perspective in the early fifteenth century preceded Galileo's use of the telescope. Likewise, Niepce's first photograph of a rooftop in Paris and Daguerre's of a Paris street precede the development of microscopy, astrophotography, and all other scientific and technological applications of photography. Likewise, the Lumiere brothers' and Edison's motion picture cameras introduced a new art form and invented the possibilities of a multitude of scientific and industrial applications. STEAM is nothing new. If the idea of integration it represents has been lost, it may only be so in our silo-ized educational institutions increasingly pressured to justify shrinking funding with immediate, quantifiable results.

At the same time, artists and scientists, engineers, and artists and scholars in other disciplines have a long history of collaboration. The Media Lab at the Massachusetts Institute of Technology, supporting collaboration between artists and scientists, opened in 1980. This is not a new idea. As art develops in close connection to the modes of economic production, various modes of art have become increasingly technical and technological. This has certainly been the case in my field of film and video. Although I cannot provide even a cursory summation of these developments, as mass media, now including web and social media, have developed as means of knowing and communicating, art has adapted and at times enhanced the development of these technologies. The transition from analogical representation to digital technologies has perhaps even been a more profound challenge in the arts than other spheres, although it has recently become evident everywhere that objective truth claims are losing their potency. One simple illustration of the degree to which contemporary media art practice overlaps with other areas such as mathematics is the relatively recent name change of the Videoformes Media Art Festival in Clermont-Ferrand, France, in existence since 1986, to the Videoformes *Numerical Arts Festival*, conveying the ever-increasing connection between media art and digital technologies and the mathematical principles of computer science.

As much as art has always been closely connected to technology, real-world practical applications have never precluded the possibility of artistry. Timothy O'Sullivan's photographs of the American west as the official photographer of the United States Geological Exploration of the Fortieth Parallel, 1867–1869, were the NASA moon landing and Mars rover photographs of their day. Yet for some time now, O'Sullivan's photographs have been recognized as some of the most accomplished and artistic uses of photography, exceeding their intended utility, with prices at auction to prove it. There are many such examples.

As I mentioned earlier, what is at risk in the current effort to reintegrate these areas is losing sight of some of the ways that they are different. Here again I am speaking about art. I may be guilty of assuming without fuller understanding that the sciences are, in general, more practically minded than art. Even if those assumptions are wrong, however, it seems that art is the domain that is being pulled into the STEM array to serve as an adjunct to those other fields, at the risk of sacrificing its other potentials, not the other way around, placing science, math, and technology in service to art. Although the tendencies within the highly heterogeneous domain of "art" are varied and fluid, there has been a relatively consistent idea that art can offer a critical perspective on the underlying assumptions that define social reality, especially since the advent of industrialism. As technologies of industrial reproduction developed, those same technologies have been used, and misused, in various ways by artists that challenge their prevalent, dominant uses underpinning social reality. The spectrum of forms of this heterogeneous activity is very broad, from benign-seeming independent play with or celebration of those technologies, sometimes as a means to humanize and integrate those technologies into the social formation, mitigating their negative effects, to iterations of the idea of "distanciation" or "making strange" the conventional realism and transparency of those technologies underpinning social reality, to outright, direct confrontation and criticism, with possibly even more subtle variations.[1] What they have in common, however, is a degree of independence from the practical utility of those technologies. Their power as criticism and confrontation, critical reflection, and even momentary escape stems from their use in non-practical ways.

It is certainly beyond the limited scope of this essay as well as my intellectual abilities in general to adequately summarize the long-standing and contentious debates concerning the nature of art and its social functions. It is tempting to give in to the commonplace idea that art is in the eye of the beholder, meaning that it is entirely up for grabs, lacking any coherent meaning beyond the personal. However, to do so also permits art to be made to mean what anyone wants it to mean and to serve any social, political, and/or economic purpose. One thinks of the effort under National Socialism in Germany as well as under Stalinism in the USSR, to elevate as propaganda a certain visual aesthetic that expressed these respective ideologies, and to eradicate other aesthetic strains labeled as decadent, as contrary to the social good. This is also a blatant admission that some forms of art serve as a thorn in the side of totalitarian regimes, potentially undermining their ideologies to the degree that they must be forcefully repressed. Of all the many possible ways to think about "art," without arguing that this one possibility is the truest or most important way, it seems to me that the idea that art can serve as a critical mirror to a social order is the one way that seems most at risk as a component of STEAM as a means of facilitating integrated learning and propelling the so-called innovation economy.

There is a long history of ways of thinking about art, vastly different ways from the Romantic to the Marxist, that nevertheless separate "art" into two main categories or planes, one serving practical utilitarian goals, and one that does not, standing outside of the practical sphere. According to Jeanne Willette, the eighteenth-century philosopher Emmanuel Kant introduced the idea of art as a purposefully *purposeless* activity:

> Kant introduces purposiveness without a purpose, allowing the mind of the one who contemplates art freely thanks to an unrestricted play of the mental faculties [...] Kant set art

[1] In an earlier essay (Arnold 1998), I examined the function of technological integration in the first two *Terminator* films (James Cameron) where fears of job loss – termination – due to automation were exploited and simultaneously displaced into the imaginary realm, as we are saved from extermination in a war with machines by a humanized robot called a "terminator."

> free from content, subject matter, the client's wishes, the community's desires and the needs of religion. The idea of art being given wholly over to aesthetic pleasure and delight was the ultimate freedom of art to exist on its own merits and to be the center of its own world. Art lived and died by its own art rules and justified its own existence in terms of its separate universe. Art was autonomous and free. (Willette 2010)

The autonomy of art, separate from other practical spheres of activity, fulfills its unique potential as a form of human experience and understanding. According to Jennifer A. McMahon, Kant identifies the unique way that art permits the human mind to process experience of the world freely rather than determined by the interests and needs of our primary physical natures (McMahon 2011). Almost a century after Kant, Oscar Wilde echoed these sentiments saying that art is "useless as a flower is useless" as an object of aesthetic contemplation free of other practical considerations (Wilde 1891). The autonomy of art, irreducible to other external – notably utilitarian – values, underpins Western art of the nineteenth and twentieth centuries, according to art historian Annemarie Bucher (2010). That autonomy, Bucher explains, is essential to art's increasingly important potential role in modern technological society as a form of social criticism – as a "protest against reality" as she puts it (Bucher 2010).

Whereas Romantic notions of "art for art's sake" and "art speaks for itself," suggesting that an art object expresses its meaning and/or value entirely pro se, and even later *New Critical* formulations that meaning is internal to the artistic "text" were rightly called into question by various *poststructural* methodologies that placed art within a broader context of social practices, the idea or possibility that art, even if understood as fully embedded within a complex and multilayered social formation, can nevertheless offer critical "distance" from within and reflection upon that social formation has remained a viable consideration. The Marxist theorist Louis Althusser placed art alongside science as a coequal branch of knowledge in the modern world in its ability to make the ideology from whence it comes visible, to "make us see" not the objects or places depicted in visual art, but the relations that bind objects, places, and times. Art is inevitably a product of ideology, but art has the unique ability to make ideology itself a subject of knowledge by virtue of an "internal distance" it produces.[2] Althusser's ideas are well summarized by Ana Ferreira as follows:

> What becomes essential then, is to be aware of the fact that categories and concepts through which we think the real are not themselves the same as immediate reality. They are mediated by operations of association, worked out in and framed by the social institutions and the state apparatuses, whose levels of operation and the way in which they induce responses it is crucial to understand. (Monteiro-Ferreira 2012)

The idea of "distanciation" is not unique to Althusser's thinking although it is frequently associated with Marxism. Perhaps its most popular proponent in the West was the playwright Berthold Brecht, whose concept of "epic theater" attempted

[2]These ideas are sketched out in two essays, "A Letter on Art in Reply to Andre Daspre" and "Cremonini, Painter of the Abstract," included in ALTHUSSER, L. 1971. *Lenin and Philosophy and Other Essays,* New York, Monthly Review Press.

to put ideas of distanciation and defamiliarization into aesthetic practice. These ideas were later picked up and developed by filmmakers such as Jean-Luc Godard. Simply put, distanciation permits the spectator to see through the cracks art can open up in the seamless façade of apparent reality that is ideology. These ideas were also central to the thinking of the Russian formalist group of literary theorists in the 1920s who attempted to systematically understand the "aesthetic effect" proper to the arts. One pivotal idea they proposed was that the function of poetry was to stand in opposition to the over-familiarization of language that is a natural consequence of its practical usage in human affairs. In quotidian usage, language tends toward transparency, and the material qualities of speech are rendered less and less perceptible, for intended meaning to be clearer. In the words of Viktor Shklovsky, the function of poetry was to "make the stone stony again":

> Habitualization devours work, clothes, furniture, one's wife, and the fear of war. "If the whole complex lives of many people go on unconsciously, then such lives are as if they had never been." And art exists that one may recover the sensation of life; it exists to make one feel things, to make the stone stony. The purpose of art is to impart the sensation of things as they are perceived and not as they are known. The technique of art is to make objects "unfamiliar," to make forms difficult, to increase the difficulty and length of perception because the process of perception is an aesthetic end in itself and must be prolonged. (Shklovsky 1965 [1917])

Like "the fear of war," we have become increasingly habituated to violence through the overuse of words, such as "terrorism," that describe it, and less aware of the complex causes that produce it, as these words gradually lose their connection to reality or supplant reality with abstraction. On a more mundane level, closer to the point, well-intended initiatives such as STEAM quickly devolve into slogans and catchphrases, such as "creative problem solving" and "the innovation economy," that lose contact with concrete meaning the more they are used.[3] Without asserting that it is the only role of art, the potential of art to defamiliarize, common (with some differences) among Shklovsky, Althusser, and many others, serves to elevate habitual, codified perception to *apperception*, permitting us to see, among many possibilities, that despite widespread belief in the assertion to the contrary, the emperor is wearing no clothes.

At this point, no one can really argue that art will undo ideology and bring about a utopian revolution. The capitalistic embrace of art as status and lifestyle commodity has been much more effective in neutralizing the socially critical potential of art than fascist or socialist repressions ever were. One need only think of the "subversive" graffiti artist Banksy, whose original work sells for a premium and is also widely available on t-shirts. This isn't to say that Banksy is a "sellout" or that his work doesn't have oppositional meaning or value – his recent self-funded dystopian theme park installation, "Dismaland," an ironic critique of fantasy simulacra such as Disneyland, is a case in point – but to illustrate the impossibility of art to exist

[3] My own video, *Echolalia* (2003), explored the deadening effect caused by the endless repetition of the phrase, "weapons of mass destruction," that served as justification for the second war in Iraq that, despite the verified absence of such weapons, still continues today.

entirely outside its social and economic system of exchange. Even so, recent developments in the USA under President Trump, who has recently proposed eliminating all funding for the National Endowment for the Arts, the federal agency that funnels public funding to regional, state, and local arts organizations, along with other cultural institutions – amounting to a trivial 0.02% of the federal budget – suggest that totalitarian tendencies to repress artistic expression are not entirely passé. One can only hope that the more overt are efforts to repress art, the more effective art becomes as a counterforce, as has often been the case in history.

How does this relate to STEAM? My question is simply how this potential of art to work against the grain of dominant ideology, to distance us from the mechanisms that construct our prevailing understanding of reality – as a "protest against reality" as Bucher put it – or even to offer a moment of perception that is *not* devoted to practical utilitarian pursuits, fits within the STEAM concept of serving practical, primarily economic goals? Returning to Banksy's "Dismaland" project, highly engineered theme parks in the mold of Disneyland that Banksy's "Dismaland" project subverts through ironic distanciation are self-congratulatory examples of STEAM, combining sophisticated science, technology, engineering, and design to achieve not only high profits but also a high order of simulation that, according to Jean Baudrillard (1994 [1981]), conceals "the emptiness of the real." Disneyland and other parks now present themselves as models of STEAM and offer STEAM "Imagineering" classroom projects.[4]

Certainly, there is a role for "art" in STEM, to facilitate learning where useful and to broaden perspectives in collaborative initiatives. If using an abstract painting as a visual aid helps students to better understand angles, as in one frequently cited example of STEAM, that's great. But at the same time, doesn't that trivialize the abstract painting, reducing its potential function to a minimal idea of utility? Is it *only* a visual representation of angles? Is there ever a follow-up discussion of how an understanding of mathematical or other STEM concepts might illuminate our understanding of abstract art?

Perhaps the issue hinges less on what one means by "art" than what one means by "creative problem solving." Art has served many different social formations to solve a variety of problems, not all of them benign, if one thinks not only of the utilitarian design of the Nazi death camps, as a gruesome example of a technical "problem" in need of a solution, as it was discussed, but perhaps more fundamentally, the systematic aestheticization of almost all aspects of life under National Socialism serving to give the ideology of fascism a seductive form rooted in a created mythology.[5] The pioneering filmmaker Leni Riefenstahl devoted her enormously creative talents to the idealization of the Third Reich, as did other artists and architects.

Less horrific by far, but still troubling, the acclaimed aesthetic design of Apple products and the unique user experience they provide serve to maintain what many

[4] http://www.mydisneyclass.com/design-a-theme-park-attraction

[5] See, for example, Kleinman, N. K., BILL 1969. *The Dream that Was No More a Dream: The Search for Aesthetic Reality in Germany 1890–1945,* Cambridge, MA, Schenkman Publishing.

believe is an unhealthy self-reinforcing fixation on using the product, as do many social media platforms, often as a platform to maintain exposure to advertising (Bosker 2016). This of course is only a more sophisticated extension of what commercial television has always done, less horrific certainly but no less a form of social control.

More broadly, my concern stems from a growing apprehension that universities, especially underfunded public universities, would like very much to do away with art programs that exceed their efforts to quantify everything or at least to constrain those departments to measurable, practical (i.e., economic) "outcomes." This threat extends well beyond the university, evident in President Trump's budget director Mike Mulvaney's recent remarks concerning cutting funding for even low-cost social programs such as "meals on wheels" and school lunch programs that don't produce "measurable results." There are, however, values that are not quantifiable. The widespread effort to reduce everything to practical and measurable outcomes is at odds with the fundamental non-practical and non-quantifiable autonomy of the aesthetic sphere and, with it, perhaps also much of the humanities. If there is a systematic logic to these efforts, even if it is just the economic "bottom line," what is the ultimate goal? Who or what is served by a world defined by practicality and measurable results, absent "useless" ideas such as art?

Producing knowledge of ideology by making its relations visible, or even just providing moments of "disinterested" aesthetic contemplation, is not among those quantifiable outcomes. My concern is that STEAM becomes an easy way of seeming to support the arts while at the same time subordinating them to other priorities, primarily economic, at the expense of the effects art can produce resulting from its non-utilitarian autonomy. Beyond the university, STEAM seems a means of subordinating art to other, practical utilitarian practices, diminishing its reflective and critical potential. Creativity may be tasked to serve many purposes, but art in its fullest realization serves no purpose, residing outside of the domain of practical utility.

References

Althusser L (1971) Lenin and philosophy and other essays. Monthly Review Press, New York

Arnold R (1998) Termination or transformation? The "terminator" films and recent changes in the U.S. auto industry. Film Q 52:20–30

Baudrillard J (1994 [1981]) Simulacra and simulation. University of Michigan Press, Ann Arbor

Benjamin W (1969 [1936]) The work of art in the age of mechanical reproduction. In: Arendt H (ed) Illuminations. Schocken, New York, pp 217–251

Berger J (1973) Ways of seeing. Penguin Books, London

Bosker B (2016) The binge breaker: Tristan Harris believes Silicon Valley is addicting us to our phones: He's determined to make it stop [online]. The Atlantic. Available https://www.theatlantic.com/magazine/archive/2016/11/the-binge-breaker/501122/

Bucher A (2010) Why art? remarks on the cultural and social functions of art. Available http://foaflux.net/wp-content/uploads/2011/04/bucher_whyart.pdf

May GS (2015) STEM, not STEAM [Online]. insidehighered.com. Available https://www.insidehighered.com/views/2015/03/30/essay-criticizes-idea-adding-arts-push-stem-education

McMahon JA (2011) Aesthetic autonomy and praxis: art and language in Adorno and Habermas. Int J Philos Stud 19:255–175

Monteiro-Ferreira A (2012) Art and ideology. *Gaudium Sciendi* [Online]. Available http://www.ucp.pt/site/resources/documents/SCUCP/GaudiumSciendi/GaudiumSciendi_N2/N2_Artigos_AnaFerreira.pdf

Shklovsky V (1965 [1917]) Art as technique. Russian formalist criticism*: Four Essays*. University of Nebraska Press

Wilde O (1891) *Letter to Bernulf Clegg* [Online]. The Morgan Library & Museum. Available http://www.themorgan.org/collection/oscar-wilde/manuscripts-letters/36?id=1535. Accessed 26 July 2017

Willette J (2010) Kant and Art for Art's Sake [Online]. arthistoryunstuffed.com. Available http://arthistoryunstuffed.com/kant-art-for-arts-sake/

Index

© Springer International Publishing AG, part of Springer Nature 2019

A. de la Garza, C. Travis (eds.), *The STEAM Revolution*,
https://doi.org/10.1007/978-3-319-89818-6

D

E

MIX
Papier aus verantwortungsvollen Quellen
Paper from responsible sources
FSC® C105338

If you have any concerns about our products,
you can contact us on
ProductSafety@springernature.com

In case Publisher is established outside the EU,
the EU authorized representative is:
Springer Nature Customer Service Center GmbH
Europaplatz 3, 69115 Heidelberg, Germany

Printed by Libri Plureos GmbH
in Hamburg, Germany